The Egg and I

BETTY MacDONALD

PERENNIAL LIBRARY
Harper & Row, Publishers, New York
Grand Rapids, Philadelphia, St. Louis, San Francisco
London, Singapore, Sydney, Tokyo, Toronto

TO MY SISTER MARY
who has always believed that I can
do anything she puts her mind to

 For information address Harper & Row, Publishers, Inc., 10 East 53rd Street, New York, N.Y. 10022. Published simultaneously in Canada by Fitzhenry & Whiteside Limited, Toronto.

First PERENNIAL LIBRARY edition published 1987.

Library of Congress Cataloging-in-Publication Data

MacDonald, Betty Bard.
The egg and I.

1. MacDonald, Betty Bard. 2. Farm life—Washington (State) 3. Washington (State)—Biography. [1. MacDonald, Betty Bard. 2. Farmers. 3. Farm life—Wit and humor] I. Title.
CT275.M43A3 1987 979.7 [B] [92] 87-45068
ISBN 0-06-091428-9 (pbk.)

06 07 08 09 10 RRD(H) 30 29 28 27 26 25

Contents

Foreword to the Paperback Edition

Betty MacDonald, our mother, is in print again. We are delighted. Her books, *The Egg and I, The Plague and I, Anybody Can Do Anything, Onions in the Stew*, the four "Mrs. Piggle Wiggle" books and our favorite, *Nancy and Plum*, continue to delight adults and children around the world.

Her first book, *The Egg and I*, was written in 1945, forty-two years ago, at a time when, according to some men, women weren't supposed to be writers; they were supposed to stay at home, bake pies and have babies. If a woman was lucky enough to have a boy, she was lavished with presents and praise. If she was unlucky enough to give birth to a daughter ("It's only a girl"), she was rewarded with a pat on the leg, a new apron and warming phrases such as "Better luck next time." An exaggeration, we know, but it sets the scene for how unprepared we were for a woman—Betty MacDonald, our mother—becoming an overnight success and a best-selling author. We felt it was a wonderful dream that might be snatched away at any moment.

We lived on Vashon Island then, and at night we would sit around the fire trying to keep warm and talk about what we would do with the money if "The Book" sold 200 copies, maybe 400. Betty wanted a fireplace in every room and a big wide road down to the house so we wouldn't have to walk the narrow, vine-covered, slug-infested trail anymore. Instead of carrying our groceries in by knapsack one and a half miles, we could have them delivered. Don MacDonald, our father, wanted to buy a case of imported Scotch, a case of Money's mushrooms and big locks for his closets so "the girls" couldn't get to his clothes. But *no road*—Don liked his privacy. We girls wanted blast furnaces installed in every room of our big, drafty house and a charge account at the Vashon Pharmacy so we could buy lipstick and nail polish. And we *didn't* want privacy—we wanted "The Road."

Two years later, *The Egg and I* was still number one on the national best-seller list. Betty got "The Road" and the fireplaces. We girls settled for radiant heat under the floors instead of blast furnaces plus the charge account at the Vashon drug store, which was to be kept under twenty-five dollars a month (which to us seemed like twenty-five hundred). Don got his case of very old, very expensive imported Scotch, his case of mushrooms and his big locks for his closet, which unfortunately for him didn't come with keys for some strange reason. Of course, he didn't know this until the locks were firmly and securely bolted in place.

Don and "The Road" was another matter. He was sure Betty's success was jeopardizing his privacy. We began to find catalogs on the kitchen table and in the bathroom with the corners of pages turned down. Things like electric fences and spikes that came out of the ground and popped tires on cars were circled and marked "send for."

At breakfast one beautiful Saturday morning we were eating scrambled eggs—with Don's mushrooms in them—and Betty was telling Don that he had lots of privacy and that he was being a "Big Black Future" (a family expression) and that nobody knew where we lived and, besides, who would take the ferry clear over to Vashon and try to find our very hidden, very unmarked road? Practically an impossibility. Don wasn't convinced, and he had just started to respond when a baby in a stroller appeared outside the kitchen window, right in front of us. The baby had a large family behind her. They carried cameras, tapped their fingers on the window and said, "Take another bite of egg, Betty." We laughed, Betty smiled obligingly and Don shouted, "Gawd, Betty do something!" and disappeared.

The fame had begun. Don's privacy was threatened but not destroyed. Ivan Dimitri, a *Life* magazine photographer, moved in with us for a week and took hundreds of pictures of Betty and us and Don, when he could find him. Don's voice was around more than

Don was, constantly admonishing: "Don't say that!" We called him "Don't Say That" Don.

Betty shared her fame with her whole family, all of her friends and her fans. We took trips to New York and stayed at the Algonquin Hotel. We met famous people, ate at famous restaurants, wore our mother's new designer clothes, pinned and rolled up, saw Broadway shows, went to Hollywood, met movie stars and went to nightclubs, autographings, radio interviews and public appearances. Through all this, Betty, being so shy and modest, did not understand why anyone would want to meet her, let alone hear her make a speech. She always said she was a nervous, unfunny wreck who sounded just like Donald Duck. Grand, glorious and glamorous Betty kept her feet—and ours—on the ground. Summer trips to Hollywood, but home to summer jobs; spring vacations in Chicago and New York, but home to baby-sitting and after-school jobs.

If anyone asked what was the greatest thing she enjoyed about her success, she would say, "Being recognized so you can cash a check anywhere."

When she saw in print everything from *The Dredge and I* (a dull Alaskan tome) to *The Cook and I, The Fish and I, The Quilt and I*, and on and on, her comment was, "I'd rather be copied than be the copier."

She answered all of her fan mail, most of it filed under "People Who Want." Thousands of letter-writers asked her to collaborate with them on their work, as "it would be a lot funnier," "not dirty" and "more interesting"; and they told her they would pay her, too—maybe. She replied with her always gracious out: "My agents don't allow me to do collaboration work of any kind."

Examples:

Dear Mrs. MacDonald:

I have read your books and find them very cheerful. I write better of course but more sad. Send my manuscript to your agent and publisher and I'll give you half of my earnings.

Sincerely yours,
Greta Swenson

Dear Mrs. MacDonald:

For some time I have wondered about writing a book about my experience—my beauty shoppe—for the past sixteen years I have operate my shoppe in this Creole country. I hope it would bring enough money to justify my paying you to write it.

Yours truly,
Alma Quilter

There were rude, ungracious remarks: "I didn't know you were so huge." (Betty was 5'9½".) "I didn't realize you'd be soooo fat."

We are certain that if Betty were alive today, she would address the plight of the American Indian in a much different manner. We feel that she only meant to turn what was to her a frightening situation into a lighthearted encounter. Remember, she had been brought up to be a lady—one who in those days was completely unprepared to handle the problems she dealt with so blithely in *The Egg*.

Through all of the funny, ugly, sad, painful, joyful times, she always shared with and gave to her reading public. She is still sharing through her books. The ability to laugh at herself and to make others laugh and her lovely spirit of optimism are as real today as they were then.

Betty MacDonald: a unique, loving, fascinating, funny, never dull, enormously talented mother is now in print again.

March, 1987

Anne MacDonald Evans
Joan MacDonald Keil

PART ONE

Such Duty

Such duty as the subject owes the prince,
Even such a woman oweth to her husband.

—SHAKESPEARE

1

And I'll Be Happy

ALONG with teaching us that lamb must be cooked with garlic and that a lady never scratches her head or spits, my mother taught my sisters and me that it is a wife's bounden duty to see that her husband is happy in his work. "First make sure that your husband is doing the kind of work he enjoys and is best fitted for and then cheerfully accept whatever it entails. If you marry a doctor, don't whine because he doesn't keep the hours of a shoe clerk, and by the same token if you marry a shoe clerk, don't complain because he doesn't make as much money as a doctor. Be satisfied that he works regular hours," Mother told us.

According to Mother, if your husband wants to give up the banking business and polish agates for a living, let him. Help him with his agate polishing. Learn to know and to love agates (and incidentally to eat them).

"It is depressing enough for a man to know that he has to work the rest of his life without the added burden of knowing that it will be work he hates. Too many potentially great men are eating their hearts out in dull jobs because of selfish wives." And Mother had examples too. There was the Fuller Brush man who came to our house once a month and told Mother how deliriously happy he used to be raising Siberian wolves and playing the violin with a symphony orchestra until he ran afoul of and married Myrtle. The man in the

A & P vegetable department who was lilting through life as a veterinary surgeon until he married a woman who hated animals but loved vegetables. And the numerous mining men Mother and Daddy knew who were held down to uninspiring company jobs by wives who wouldn't face the financial insecurity of their husbands going into business for themselves.

"Boy," we said, "when we get married, our husbands will do exactly as they please," and they have.

This I'll-go-where-you-go-do-what-you-do-be-what-you-are-and-I'll-be-happy philosophy worked out splendidly for Mother for she followed my mining engineer father all over the United States and led a fascinating life; but not so well for me, because although I did what she told me and let Bob choose the work in which he felt he would be happiest and then plunged wholeheartedly in with him, I wound up on the Pacific Coast in the most untamed corner of the United States, with a ten-gallon keg of good whiskey, some very dirty Indians, and hundreds and hundreds of most uninteresting chickens.

Something was wrong. Either Mother skipped a chapter or there was some great lack in me, because Bob was happy in his work but I was not. I couldn't learn to love or to know chickens or Indians and, instead of enjoying living in that vast wilderness, I kept thinking: Who am I against two and a half million acres of mountains and trees? Perhaps Mother with her flair for pioneering would have enjoyed it. Perhaps.

Where Mother got this pioneer spirit, how she came by it, I do not know, for a thorough search of the family records reveals no Daniel Boones, no wagon trains heading West with brave women slapping at Indians with their sunbonnets. In fact, our family tree appears rife with lethargy, which no doubt accounts for our all living to be eighty-seven or ninety-three.

Mother's ancestors were Dutch. Ten Eyck was their name and they settled in New York in 1613. One of my father's family names was Campbell. The Campbells came to Virginia from Scotland. They were all nice well-bred people but not daring or adventuresome except for "Gammy," my father's mother, who wore her corsets upside down and her shoes on the wrong feet and married a gambler with yellow eyes. The gambler, James Bard of Bardstown, Kentucky, took his wife out West, played Faro with his money, his wife's money and even some of his company's money and then tactfully disappeared and was always spoken of as dead.

We never saw this grandfather but he influenced our lives whether he knew it or not, because Gammy was a strong believer in heredity, particularly the inheritance of bad traits, and she watched us like hawks when we were children to see if the "taint" was coming out in any of us. She hammered on my father to such an extent about his gambling blood that he would not allow us children to play cards in any form, not even Slap Jack or Old Maid, and though Mother finally forced him to learn to play double Canfield, he died without ever having played a hand of bridge, a feat which I envy heartily.

The monotony of Mother's family was not relieved in any way until she married Darsie Bard who was her brother's tutor and a *Westerner working* his way through Harvard. This was a very shocking incident as Mother's family believed that the confines of civilization ended with the boundaries of New York State and that Westerners were a lot of very vulgar people who pronounced their r's and thought they were as good as anybody. Mother's mother, whom later we were forced to call Deargrandmother, had fainting fits, spells and tantrums but to no avail. Mother went flipping off without a backward glance, to live, for Heaven's sake, in Butte, Montana.

This was Butte in the early 1900's. The time of the Copper Kings, when everyone made a million dollars, there were thirty-five thousand miners working underground and a saloon every other doorway. Irish scrubwomen became the wives of millionaires and had interior decorators come from France to "do" their houses. Lawns were imported blade by blade and given the care of orchids in order to make them grow in that sulphur-laden air. Oriental rugs were a sign of wealth and grandeur and were put on the floors three deep and piled in heaps in attics. Southern mansions, French chateaus, Welsh stone cottages, timbered English houses, Swiss chalets and American bungalows were built to house the rich Irish. Everyone was cordial, bluff and gay and entertained lavishly and all the time. A party was given at the Silver Bow Club to welcome Mother to Butte and she was amazed to find that the ladies of the town wore Paris gowns but painted their faces like prostitutes. Mother had been reared to believe that if you were unfortunate enough to be born with a pale green face, you, if you were a lady, would not for a moment entertain the thought of rouge, but would accept your color as your cross and do nice things for poor people. Mother had been reared this way but she didn't endorse such nonsense. Fortunately she had natural colors so she wasn't put to any test, but she was certainly pleased to find that the ladies of Butte, and there were many ladies in the strictest sense of the word, had kicked over the traces from Boston to Atlanta and were improving on nature with everything they could lay hand to. Mother loved the West and she loved Westerners.

My sister Mary was born in Butte. She had red hair and to appease Mother's family was given the middle name of Ten Eyck which necessitated her fighting her way through grammar school to the taunts of Mary Tin Neck.

When Mary was less than a year old my father was sent down to the Nevada desert to examine gold property. Mother joyfully went with him and lived in a shack and rode horseback with the baby on the saddle in front of her. Both Mother and Daddy were happy in his work.

I was born in Boulder, Colorado. Gammy was with us then and the night I was born, when Mother began having pains, she called to Gammy (Daddy was away on a mining trip) and told her to phone for the doctor and nurse. But Gammy, prompted by the same inner urge which made her wear her corsets upside down, rushed across the street and pounded on the door of a veterinary and when he appeared, she dragged him bewildered and in his long underwear to Mother's bedside. Mother, very calm, sent the poor man home, but because of the delay and confusion I was born before the doctor could get there and it was necessary for Gammy to tie and cut the umbilical cord. This was very unfortunate, as Gammy, a Southern girl, had been "delicately reared" and her knowledge of rudimentary anatomy could have been put in the eye of a needle. She thought the cord had to be tied into a knot and so grabbing me like the frayed end of a rope, she began looping me through and under as she attempted the knot. The upshot was that Mother sat up and tied and cut the cord herself and I was named for Gammy and became another in a long line of Anne Elizabeth Campbells. My hair was snow white but later turned red.

When I was a few months old Mother received the following wire from Daddy: "Leaving for Mexico City for two years Thursday—be ready if you want to come along." This was Monday. Mother wired: "Will be ready" and she was; and Thursday morning, we all, including Gammy, left for Mexico.

Díaz was serving his last term as president of Mexico then, and Mexico City was a delightful place of Mexicans, flowers and beautiful horses. My sister, Mary, because of her brilliant red hair, was much admired by the Mexicans and learned to speak fluent Spanish, but I, an outstanding dullard, didn't even begin to speak anything until almost three. There was a series of violent earthquakes while we were in Mexico but Mother, never one to become hysterical, inquired as to earthquake procedure, and when the lamps began to describe arcs in the air, the mirrors to sway and the walls to buckle, Mother sensibly herded Mary, Gammy and me into the doorway of the apartment, where the building structure was supposed to be strongest and, though the apartment building was cracked from top to bottom, we were all unharmed. A woman in the next apartment became very excited and rushed into the street in her nightgown, where, I am happy to relate, a water main burst directly under her.

From Mexico we moved to Placerville, Idaho, a mining camp in the mountains near Boise, where the snow was fifteen feet deep on the level in winter and Mother bought a year's supply of food at a time. Our closest neighbor was a kind woman who had been a very successful prostitute in Alaska and wore a chain of large gold nuggets which reached below her knees. She was very fond of me, Mother says, and told everyone, to Gammy's intense annoyance, that I was the "spitting image" of her when she was three years old. In Placerville, Mrs. Wooster (I believe that this was her name) had become a respectable married woman but evidently this palled, for Mother says she talked constantly of the "good old days." I can feel for her because, although I have never been an Alaskan prostitute dancing on the bar in a spangled dress, I still got very bored with washing and ironing and dishwashing and cooking day after relentless day. Of course Mrs.

Wooster had an extra hurdle in her path of boredom, that of the same old husband jumping into bed every night.

In Placerville, my father supervised his first large placer mining project and as the work was both dangerous and hard, Mother tore the partitions out of the crackerbox house, built a fireplace and bore my brother Sydney Cleveland, all by herself. Cleveland had red hair. All this red hair caused a lot of comment in Placerville, as Mother was a blonde with brown eyes and Daddy had jet black hair and gray eyes. What no one knew was that Daddy had a bright red beard if he let it grow. When Cleve was born Mother's father wired her, "I trust you won't feel called upon to have a child in every state in the Union."

Our next jaunt was East to visit Mother's mother or Deargrandmother. The moment we arrived, we children were stuffed into a nursery with an adenoidal nurse named Phyllis, and at my five-year-old birthday party the children were instructed by Deargrandmother not to bring presents. My, how we longed for Gammy with her shoes on the wrong feet and her easy friendly ways. Deargrandmother was noted for her beautiful figure and proud carriage but she toed out and had trouble with her arches. She taught Mary and me to turn our toes out when we walked, say "Very well, thank you" instead of "Fine" when people inquired of our health, and to curtsy when we said "How do you do?" She tried hard to scrape the West off these little nuggets, but as soon as we returned home Daddy made us walk like Indians again, feet pointed straight ahead. I would like to remark here and now, that this walking with feet pointed straight ahead is the only thing about an Indian which I would care to imitate.

When we returned from Auburn, we moved to Butte and lived there for the next four years.

Of Butte I remember long underwear which Gammy called

"chimaloons" for some strange reason of her own. We folded our "chimaloons" carefully at the ankles so as not to wrinkle our white stockings. I remember my new Lightning Glider sled and coasting fourteen blocks downhill on Montana Street and hitching a ride all the way back. I remember icicles as big as our legs hanging outside the windows, and bobsledding at night with Daddy, who invariably tipped the sled over and took us home bawling. Creamed codfish and baked potatoes for breakfast and hot soup with grease bubbles which Gammy called "eyes" in it for lunch. Walking to the post office with Daddy on Sunday night and holding bags of popcorn in our clumsy mittened hands and drinking the hot buttery popcorn out of the bag. The Christmas when we had scarlet fever and the thermometer went down and stuck in the bulb and we got wonderful presents which had to be burned. Creaking down the street through the dry snow to dancing school, our black patent leather slippers in a flowered bag, our breath white in front of us. A frozen cheek that Mother thawed with snow. A wonderful sleighride into the mountains at night with the bells sounding like tinkling glass, the runners hissing softly and our eyes peering from heaps of robes.

When Cleve and I used to "rassle" to see who would get the biggest apple, the most candy, or any of the other senseless things children quarrel about, Gammy would stand over us and shout, "Get the hatchet, Cleve, and kill her now. You'll do it some day, so why not now." This infuriated us so that we would cease pounding each other and become bosom friends just to "show" Gammy. Perhaps this was her underlying motive but it used to seem to me that she was far too anxious to get rid of her little namesake.

My sister Darsie was born when I was in the second grade. She was small and had dark hair. Also in the second grade a little boy named Waldo wet his panties while we were standing in the front of the class for reading and I got so red in the face that the teacher, a horrid creature who said "wite" for white and "tred" for thread, blamed me and felt my panties, to see if they were dry, in front of the whole class.

Mary and I wore white stockings to school every day and shoes with patent leather bottoms and white kid tops. Mary turned her stockings wrong side out and wore them two days which would have been all right but she told everybody and I was ashamed. Gammy made us wear aprons which she called "aperns" over our dresses while we were playing after school. She would greet us at the door with the "aperns" but if we managed to sneak out without them she would stand on the porch and call in a high mournful wail, "Giiiiiiiirls, come get your 'aperns' " (this last a high banshee shriek). After school, if the weather was nice, we played on the Montana School of Mines dump and found lots of the little clay retort cups in which gold had been assayed.

When we said bad words, which we did as fast as we learned them, and Gammy or Mother learned of it, we were given "heart medicine." This was a dark vile-tasting liquid which while shriveling our tongues was supposed to be purifying our hearts. I learned later that it was bitter cascara and no doubt served a double purpose. We could not understand why our "hired girls," who said God and Jesus all the time, were never given "heart medicine" while we innocent little children used up about a bottle a week.

Our "hired girls" were hot tempered Irish girls who hated children, especially children with red hair, and smacked us and threatened to quit if we came into the kitchen. They showed surprising weaknesses, however, like the Mary whom Mother found one frosty morning weeping into the hotcake batter. "What is the matter?" Mother asked, thinking it was probably a man. "Jesus-God, Mrs. Bard, I can't get the damned things round," and she tearfully pointed out a heap of oblong and oval hotcakes she had thrown into the sink.

Butte had no budding trees, no spring flowers and no green grass, but we knew when spring came by the raging torrents that ran in the gutters. In one such torrent I found a five-dollar bill. I thought it was a shoe coupon—I was collecting them—and carefully scooped it out with the toe of my rubber and took it home to Gammy, who ironed it dry and told me that it was five dollars. This was the first paper money I had ever seen, as silver and gold were used exclusively in Butte, and I didn't really feel that I had found five dollars until Daddy exchanged it for a gold piece which I put in my copper bank with the picture of the Anaconda Smelter on the front of it. Later Cleve and I hacked this little bank open with a mining pick and spent the five-dollar gold piece on penny candy.

In the spring Gammy took us for walks in the hills and we were careful not to fall in "prospect holes" which suddenly appeared at our feet, black, scary and bottomless. Gammy told us stories of heedless children who scampered off into the hills to play but never came back and years and years later their little white skeletons were found in "prospect holes." We gathered bluebells and bitterroot daisies and wild garlic. The bluebells were a deep clear blue like fallen sky against the bare black rocks. The bitterroot daisies, the Montana State flower, had little foliage and no stems and lay flat and pink and exquisite on the brown hard earth. We painstakingly dug them up, careful of the roots, carried them home and planted them and they immediately died as all the topsoil of Butte was washed away years before in the placer mining days, and our yard was nothing but decomposed granite. We had one patch of grass in our front yard about the size of a pocket handkerchief. I played there with my dolls, but was very careful not to sit on the grass or injure it in any way. (What would a life-long resident of Butte think if he could have seen the country surrounding our chicken ranch—where fenceposts sprouted, vines crept into the house and everything was so green, green, green, it made me feel bilious?)

One winter in Butte, we were taken to see a play at the Broadway Theatre. The play was *The Bird of Paradise* and we all clung to Gammy's hands and bawled when the beautiful heroine threw herself into the erupting volcano. The next spring we climbed Big Butte, a bare, brown mountain, a thousand feet high and almost in our backyard and were horrified on reaching the top to find a large crater and to have Gammy explain casually that this mountain, the very one on which we lay panting, was a volcano. We ran every inch of the way home, peering back over our shoulders expecting to see the top of poor old Big Butte a fiery furnace with white hot lava oozing down its sides. We never would go up there again and when the sulphur smoke hung low over the city, veiling the top of Big Butte, we were sure it was erupting.

The sulphur smoke smelled awful but Gammy made us breathe deep and suck it down inside of us. She said it disinfected our insides. She also made us drink gallons of vile-tasting water at White Sulphur Springs. Between the "heart medicine" and the sulphur water and smoke we should have been as pure as angels, but unfortunately this was not the case, for we worked diligently to find out where babies came from, until one fateful day when my sister, Mary, proclaimed to the assembled neighborhood children, from a little platform we had erected in the backyard, "Ladies and Gentlemen, babies are born out of people's stomach holes." I can still taste the heart medicine.

Gammy used to walk us downtown but as she made us close our eyes every time we passed a saloon the walks were valuable to us more for the fresh air than for the sights we saw. Once she had us open our eyes to see a hat in Hennessy's store window which cost $105. We could not get over it. One hundred and five dollars for a hat! We made three trips to see that hat but I haven't the faintest recollection of what it looked like, no doubt because I kept my eyes glued to the price mark. We had heard from Mother that Hennessy's, the company store, also sold Paris gowns but they didn't put these in the window so we never saw one.

Often cowboys in chaps and ten-gallon hats rode cayuses down Main Street and several times Indian braves on ponies, followed by squaws on foot with papooses on their backs, filed slowly past the one-hundred-and-five-dollar hat window. These were the Blackfeet Indians and they wore beautifully beaded dresses and chaps and terrific feather headdresses, and had long noses and cold Indian eyes. As Gammy had read us the stories of Hiawatha, Pocahontas and Sitting Bull and told us many hair-raising tales of massacres, scalpings and running the gantlet, we thought these Indians were simply wonderful, so strong and brave, and would run for blocks to see them. I still harbored these romantic notions about Indians when I moved to the chicken ranch, and it was a bitter blow when I learned that today's little red brother, or at least the Pacific Coast variety which I saw, is not a tall copper-colored brave, who, clad only in beads and feathers and brandishing a bow and arrow, bounds around in the deep woods. Instead, our Indian, squat and mud-colored, was more apt to be found slouched in a Model T, a toothpick clenched between his yellow teeth, a drunken leer on his flat face. On the reservation he was orderly and well behaved and, we were told, used to engage in dangerous pursuits like whaling and seal hunting; but in appearance, at any rate, he resembled the story-book variety and my childhood Blackfeet Indian, about as much as a mud shark resembles a Beardsley trout.

Our summers were spent camping in the mountains. Usually we had a camp man and slept in tents and followed Daddy about while he examined mines, but other times we had cabins on a lake and stayed with Gammy while Mother and Daddy did the travelling. My still-smoldering hatred for and distrust of wild animals were implanted on these camping trips. Once we almost fell on a large bear, placidly eating huckleberries on the other side of a log. Another time Daddy pointed out a mountain lion lying in the sun on a ledge above our heads. Bears were always knocking down our tents and eating our supplies and at night the coyotes and timber wolves howled dismally.

Mother and Daddy fished incessantly and we had Rainbow trout, which we children loathed, three times a day. Sometimes Gammy came camping with us but only when we had cabins and didn't spend our days "traipsing" through the mountains. Gammy stayed with us while Mother and Daddy took trips and fished and although they were considerate and always asked if we cared to come along, we always refused because Mother and Daddy loved danger and were always walking logs over deep terrible ravines; walking into black dangerous mine tunnels; wading into swift turbulent streams and doing other scary things. Gammy, on the other hand, carefully avoided danger and was constantly on the alert for it.

Summer days with Gammy were spent in her cabin with the doors and windows shut tight against the dangers of mountain air. We would all crouch around her rocking chair while she read to us out of *Pilgrim's Progress* and fed us licorice drops out of her black bag. This routine was varied occasionally by a thunderstorm, whose first clap of thunder sent us hurtling under the bed clutching feather pillows and praying, or by very short walks during which Gammy called us all to a halt every few feet to listen for rattlesnakes. She had us whipped into a state where the rattle of a leaf would turn us white and sweaty and send us scurrying home to the safety of the cabin. Gammy impressed us with all of the dangers of outdoor living. She warned us against eagles, hawks, bees, flies—horseflies which bit—mosquitoes and gnats which might attack from the air; ticks, snakes, leeches, and bugs which might spring snarling from the ground; and she had us convinced that the trees along the edge of the clearing where our cabins were, were like the bars of the cages at the zoo and just behind them prowled hundreds of timber wolves, grizzly bears and mountain lions fighting for a chance to eat us.

From the summers we spent with Mother and Daddy camping in tents, we returned to town brown and healthy, but from the summers spent with Gammy, we came back as jumpy as fleas and pale and scraggly from the hours of lying on feather pillows under the beds praying during the thunderstorms and the days crowded in the close cabins out of the reach of groping fangs. We, of course, never told brave, fearless Mother and Daddy about Gammy and the dangers of outdoor life, and they probably wondered why they, so strong and daring, should have produced this group of high-tensioned rabbits.

When Mother and Daddy went away from home on long trips, which they did frequently, we stayed at home with Gammy. She had us all sleep in her room on army cots and folding beds which she hastily and carelessly erected and which were always collapsing and giving us skinned noses and black eyes. Gammy kept a pair of Daddy's shoes beside her bed and when she heard any noise in the house she leaned out of bed and stamped the shoes on the floor so that the robber or killer, whichever one happened to be downstairs, would think that there was a man in the house instead of "a lone helpless woman and several small children" all huddled upstairs waiting to be killed.

Our "hired girls" often came in late and I've wondered since if this stamping of manly feet upstairs in the dead of night, when they knew that Mother and Daddy were in New York or Alaska, didn't lead them to believe that Gammy had a secret love life. To the casual eyes of a maid this idea might have been plausible, as Gammy was a very pretty woman, small with large blue eyes, delicate regular features and tinselly curly hair. But to those of us who knew her there were several good reasons why this wouldn't, couldn't be. In the first place Gammy hated men—all men, except Daddy. "Just like some big stinkin' Man," she would sneer as she lapped up the account of a rape or murder in the paper. Or, "The whole world's run for Men and don't you forget it," she would warn us as she inspected us to see if our eyes were shut before marching us past the Silver Dollar Saloon. Or, when we were having mining men, friends of Daddy's, to dinner, which we did six nights a week, Gammy would caution the hired girl, "Don't make it so awful good. Men'll eat anything. The pigs!"

In the second place, any lover of Gammy's would have had to equip himself with enduring desire and a bowie knife, for Gammy was well covered. She thought nakedness was a sin and warned us, "Don't let me catch you running around in your naked strip!" and for her own part, she merely added or removed layers of clothing as the weather demanded. On top she always had a clean, ruffly white "apern"—during the day this was covered by a large checked "apern." Under the aprons were a black silk dress, a black wool skirt, a white batiste blouse with a high collar, any number of flannel petticoats, a corset cover, the upside-down corset with the bust part fitting snugly over the hips, and at long last the "chimaloons."

In the third place, a lover of Gammy's certainly would have had a lumpy couch with her nightgowns, bed jackets and several extra suits of "chimaloons" folded under the pillow, her Bible tucked under the sheet at the top right-hand side, any book she happened to be reading tucked under the sheet on the other side, little bags of candy, an apple or two, current magazines, numerous sachets and her bottle of camphor just tucked under the blankets or scattered under the pillows within easy reach. We children thought this an ideal arrangement, for when we were lonely or frightened Gammy's bed was as comforting as a crowded country store.

Gammy was an inexhaustible reader-aloud and took us through the Bible, *Pilgrim's Progress,* Dickens, Thackeray, Lewis Carroll, Kipling, *The Little Colonel, The Wizard of Oz, The Five Little Peppers,* and all of Zane Grey, which we adored, before we left Butte. She changed long words to ones we could understand without faltering, but after an hour or two with *The Little Colonel* or *The Five Little Peppers* she would begin to doze and we would be dispatched to the kitchen to ask Mary the Cook for some black coffee. Usually this revived her completely and she would continue until lunch or supper or bedtime, but sometimes, especially during the nauseous antics of the Little Colonel or the continual bawling of the Five Little Peppers who cried when they were happy, Gammy would drink cup after cup of black coffee but would still fall asleep and when she awoke would read the same paragraph over and over. We would make several futile trys to wake her and then would give up and go out to play.

Gammy was patient, impatient, kind, caustic, witty, sad, wise, foolish, superstitious, religious, prejudiced and dear. She was, in short, a grandmother who is, after all, a woman whose inconsistencies have sharpened with use. I have no patience with women who complain because their mothers or their husbands' mothers have to live with them. To my prejudiced eye, a child's life without a grandparent *en residence* would be a barren thing.

2

Battre L'Eau avec un Bâton

WHEN I was nine years old we moved to Seattle, Washington, and the pioneering days were over and preparedness for the future began. At least I'm quite sure that is what Mother and Daddy had in mind when they started Mary and me taking singing, piano, folk dancing, ballet, French and dramatic lessons. If they had only known what the future held, at least for me, they could have saved themselves a lot of money and effort because for my life on the chicken ranch a few hours a day shut in the icebox contemplating a pan of eggs would have been incalculably more useful early training than, say, French or ballet. French did come in handy in reading books by bilingual Englishmen and women, but conversationally it was a washout, as I did most of my talking to myself, and only Frenchmen go around talking to themselves in French.

In addition to our injections of culture, we children were suddenly tumbled into a great health program. We ate no salt, never drank water with our meals, chewed our food one hundred times, got up at five o'clock in the morning and took cold baths, exercised to music and played tennis. Also, to keep our minds healthy, I guess, we were not allowed to go to the movies or to read the funny papers. One of the houses we lived in had belonged to the Danish Consul and had a large ballroom in the basement which Daddy immediately turned into a gymnasium with horizontal bars, basketball hoops and mattresses. Every night he forced us into this torture chamber for a workout. We leaped over the bar without hands, swung by our knees, played basketball, did back flips and hated Daddy. We did not want to be healthy. We wanted to go to the movies, read the funny papers and relax like all the other unhealthy children we knew. Fortunately Daddy left home on mining trips quite often and the moment the front door closed on his tweed-covered back we got out several months' supply of funny papers and settled down to a life of hot baths and blissful slothfulness until he returned. His mining trips kept him away from home about six months of the year, off and on, and it is a wonder that our muscles withstood this business of being hardened up like flints, only to squash back to jelly. Only the lessons kept on while Daddy was away, as Mother and Gammy weren't any more anxious to get up at five o'clock and take cold baths and exercises than we were.

I have been told that I was directly responsible for this dreadful health complex of Daddy, for I was a thin, greenish child who caught everything. Up to this time I had brought home and we had all had measles, both German and Allied, mumps, chickenpox, pink eye, scarlet fever, whooping cough, lice and the itch. Every morning before sending me off to school, Mother and Gammy would examine me in a strong light to see what I had broken out with during the night, for I looked so unhealthy all of the time that they were unable to determine if I were coming down with a disease until the spots appeared.

We always lived in large houses because Daddy had a penchant for inviting people to stay with us. He would casually wire Mother from Alaska "Meet the *SS Alameda* on Thursday—Bill Swift and family coming to Seattle for a few months—have asked them to stay with you." Mother would change the sheets on the guest-room beds, heave a sigh and drive down to meet the boat. Sometimes Bill Swift and his wife and children were charming and we regretted to see them go, but other times Bill Swift was the world's biggest bore, his wife whined all the time and we fought to the death with the children. After the first day, we could tell what the guests were like from Gammy, for if they were interesting, charming people Gammy retaliated in kind and was her most fascinating and witty self, but if they were dull or irritating in any way, Gammy would give us the signal by calling them all by wrong names. If the name was Swift, Gammy would call them Smith, Sharp or Wolf. If one of the children was Gladys, Gammy would call her Gertrude or Glessa, and a boy named Tom would become Tawm. Gammy had other subtle ways of letting them know they were in the way. From her bedroom on the second floor she would call to us children playing in the basement, "Cheeldrun, please come up and see if those bores are still in the bathroom. I've been waiting an hour to get in." We thought this very clever of Gammy as we knew that the guests knew that there were other bathrooms and we would look knowingly at each other and giggle and let her call about five times before we answered. Mother must have drawn heavily on her wealth of charm and tact during those days, for in spite of Gammy's remarks all of our guests stayed their full time and all seemed sorry to go.

When I was eleven and just about ready to go up on my toes in ballet, we bought a house in Laurelhurst near the water. This was a fine big place with an orchard, a vegetable garden, tennis courts and a large level lawn for croquet. We immediately bought a cow (which obligingly had a calf), two riding horses, two dogs, three cats, a turtle, white mice, twelve chickens, two Mallard ducks, several goldfish and a canary. Our animals were not very useful and too friendly and hovered in the vicinity of the back porches day and night. We had a schoolboy who milked the cow, fed the calf, curried the horses and tethered them all out, but either he was weak or they were strong, for the minute he left for school they would all come galloping home to the back porches where Gammy fed them leftover batter cakes, toast and cocoa. We loved all of our animals and apparently our guests did too, or if they didn't love them they didn't mind them, for our house overflowed with guests and animals all of the time. Guests of Daddy's, guests of Mother's, guests of Gammy's, and our friends and animals. There were seven of us, counting Daddy who was rarely home, but our table was always set for twelve and sometimes forty. Dinner was an exciting event and we washed our knees, changed our clothes and brushed our hair with anticipatory fervor. Mother sat at one end of the table and Daddy at the other, if he was home, Gammy sat at Daddy's right and we children were spaced to eliminate fighting. Daddy had made a rule and it was strictly enforced, whether or not he was home, that only subjects of general interest were to be discussed at the table. This eliminated all such contributions from us, as "There is a boy in my room at school who eats flies," and "Myrna Hepplewaite stuck out her tongue at me and I said bah, bah, bah and she hit me back and I told her mother. . . ." In fact, it precluded our entering the conversation at all except on rare occasions which I think was and is an excellent idea. I resent heartily dining at someone's house and having all my best stories interrupted by "Not such a big bite, Hubert," or "Mummy, didn't you say the Easter Bunny came down the chimney?"

As soon as we were settled in Laurelhurst, Daddy decided that in addition to Mary's, Darsie's and my singing, piano, ballet, folk dancing, French and dramatics and Cleve's clarinet lessons, we should all have lessons in general usefulness and self-reliance. His first step in this direction was to have Mary and Cleve and me paint the roof of our three-story house. The roof was to be red and we were each given a bucket of paint, a wide brush, a ladder and some vague general instructions about painting. It seems that there was a shortage of ladders so Cleve and I were on the same one—he was just a rung or two ahead of me and both of us biting our lips and dipping our brushes and slapping on the red paint for all we were worth. We weren't working hard because we liked this job; we didn't, we just thought it was another one of Daddy's damnfool notions and we wanted to get it over with as quickly as possible. Cleve and I had just finished the small area over the back porch and were moving up when something went wrong and Cleve dumped his bucket of paint over my head and down the back of my neck. Gammy cleaned me off with turpentine but she grumbled about it and said, "It's a wonder to me you aren't all dead with the ideas some Men get." We finished the roof, though, with Daddy lowering me by the heels so that I could paint the dormers of the attic, but it was a scary, slippery job and was an outstanding failure as far as a lesson in self-reliance was concerned. Daddy's next step was the purchase of a .22 rifle and a huge target. Gammy had hysterics. *"Guns,"* she bawled. "Guns are for Huns and heathens. Those children will kill each other—please, Darsie, don't give them a gun." So we learned to shoot. Mary and I were both rather nearsighted and very poor shots but Cleve was a good shot and practised all the time. Cleve became such an expert marksman that he took up hunting when he was only ten years old and Daddy thought it was a fine idea until Cleve drew a bead and fired at a quail that was perched on the sill of a huge curved bay window of a neighbor's house. None of the neighbors was killed but the bay window was very expensive and so the gun was put away for a while and Daddy bought us an enormous bow and arrow and a big straw target. While he and Cleve were practising archery, Mary and I were learning to cook. Mother supervised this herself as she was a marvellous cook and Gammy was the world's worst. Mother taught us to put a pinch of clove and lots of onion in with a pot roast; to make French dressing with olive oil and to rub the bowl with garlic; to make mayonnaise and Thousand Island dressing; to cook a sliver of onion with string beans; never to mash potatoes until just before serving; to measure the ingredients for coffee; and always to scald out the teapot.

Gammy taught us that when you bake a cake you put in anything you can lay your hands on. A little onion, several old jars of jam, leftover batter cake dough, the rest of the syrup in the jug, a few grapes, cherries, raisins, plums or dates, and always to use drippings instead of butter or shortening. Her cakes were simply dreadful—heavy and tan and full of seeds and pits. She made a great show of having her feelings hurt if we didn't eat these cakes but I really think she only offered them to us as a sort of character test because if we were strong and refused, she'd throw them out to the dogs or chickens without a qualm.

Gammy said she did not believe in waste and she nearly drove our maids crazy by filling up the icebox with little dishes containing one pea, three string beans, a quarter of a teaspoonful of jam or a slightly used slice of lemon. If Mother finally demanded a cleanup and began jerking dishes out of the refrigerator and throwing stuff away, Gammy would become very huffy and go out and get a twenty-five pound sack of flour and hand it to Mother, saying, "Go on, throw this away too. Waste seems to be the order of the day." Gammy made great big terrible cookies, too. Into these she put the same ingredients she put in the cakes but added much more flour. These cookies were big and round and about half an inch thick. They stuck to the roof of the mouth and had no taste. What to do with them became quite a problem when we finally settled down and weren't moving around any more. They were stacking up alarmingly in the kitchen and lying around the back porch untouched when the Warrens moved across the street from us. The Warrens had a beautiful colonial house and two cars, but their children—there were four of them, two boys and two girls—ate dog biscuits. Why, I don't know, but they did. Mrs. Warren kept a one-hundred-pound sack on the back porch and the little Warrens filled their pockets after school and nibbled at them while playing Kick the Can. We tried some once, and they weren't much of a shock after Gammy's cakes but we didn't care for the rather bitter tang they had—it was no doubt the dried blood and bone. One day the Warren children stopped at our house before going home for their dog biscuits and Gammy happened to be baking cookies (she happened to be baking cookies about six days a week—she said that they were cheap and filling and would save on the grocery bill) and she forced us all to take some. The Warrens liked them. We were amazed and took a few tentative bites ourselves to see if these cookies might be different. But they weren't. They were the same big stuffy, tasteless things they had always been, but I guess compared to dog biscuit they were delicious because the Warrens begged for more and the suckers got them. All they could eat and all we couldn't eat. From that day on they ate all Gammy's output and we didn't have to flinch as we watched her pour the rest of the French dressing and a jar of "working" plums into her cookie dough.

When I was twelve years old Daddy died in Butte of streptococcic pneumonia. My sister Alison, who has red hair, was born five months later. It was a very sad year but rendered less tragic and more hectic by a visit from Deargrandmother, who came out to comfort Mother and make our lives a living hell. She dressed Mary and me in dimities and leghorn hats; asked *who* our friends were and *what* their fathers did; she wouldn't let Gammy work in her garden as it was unbecoming to a lady, so Gammy had to sneak out and hoe her potatoes and squash at eleven o'clock at night; she wouldn't let our old Scotch nurse eat at the table with us and insulted her by calling her a servant; she picked her way downtown as though we had wooden sidewalks; and was "amused" by anything she saw in our shops because this wasn't New York. Our only recourse was to go out to the laundry, which was a large room built on the back of the house and connected to the kitchen by a series of hallways and screened porches, with Nurse and Gammy, where we would make tea on the laundry stove and talk about Deargrandmother.

When she finally left for New York we took life in our own hands again and things continued much the same as they had before Daddy died except we were poorer and fewer of our guests were Mother's and Daddy's and Gammy's friends and more and more of them were friends of Mary's. As an economy measure we had stopped all our lessons but the piano and the ballet, and we were to go to public schools in the fall.

In high school and college my sister Mary was very popular with the boys, but I had braces on my teeth and got high marks. While Mary went swishing off to parties, I stayed home with Gammy and studied Ancient History or played Carom or Mahjong with Cleve. Mary brought hundreds of boys to the house but she also brought hundreds of other girls, so I usually baked the waffles and washed the dishes with a large "apern" tied over my Honor Society Pin and my aching heart. Gammy used to tell me that I was the type who would appeal to "older men," but as my idea of an older man was one of the Smith Brothers on the coughdrop box I took small comfort in this. To make matters worse I suddenly stopped being green and skinny and became rosy and fat. I grew a large, firm bust and a large, firm stomach and that was not the style. The style was my best friend, who was five feet ten inches tall and weighed ninety-two pounds. She had a small head and narrow shoulders and probably looked like a thermometer, but I thought she was simply exquisite. I bought my dresses so tight I had to ease into them like bolster covers and I took up smoking and drinking black coffee but still I had a large, firm bust, just under my chin, and a large, firm stomach slightly lower down. I am sure that Mary also had a bust and stomach but hers didn't seem to hamper her as mine did me. Perhaps it was because she had "life." "Torchy" they called her and put under her picture in the school annual: "Torchy's the girl who put the pep in pepper." Under my picture was printed in evident desperation "An honor roll student—a true friend."

I was handy around the house and Mother taught me to mitre sheets at the corners and to make a bed as smooth as glass. Gammy smoothed up her beds right over cold hot-water bottles, books, toys, nightgowns or anything else that was dumped there in the hurry of the morning. Mother insisted that anything worth doing is worth doing well, but Gammy said, "Don't be so finicky. You'll just have to do it over again tomorrow."

Mother set the table with candles and silver and glassware and flowers every night whether we had company or not. Gammy preferred to eat in the kitchen with peeler knives and carving forks as utensils. Mother taught me to wash dishes, first the glassware, then the silver, then the china and last the pots and pans. Gammy washed dishes, first a glass, then a greasy frying pan, then a piece or two of silver. Mother served food beautifully with parsley and paprika and attractive color combinations of vegetables. Gammy tossed things on the table in the dishes in which they had been cooked and when she served she crowded the food into one frightened group, leaving most of the plate bare. "After all it's only nourishment for the body," she would say as she slapped a spoon of mashed potato on top of the chop and sprinkled the whole thing with peas. It was a lesson in cross-purposes and the result now is that one day I barely clean my house and the next day I'm liable to lick the rafters and clean out nail holes with a needle.

When I was seventeen years old and a sophomore in college, my brother, Cleve, brought home for the weekend a very tall, very handsome older man. His brown skin, brown hair, blue eyes, white teeth, husky voice and kindly, gentle way were attributes enough in themselves and produced spasms of admiration from Mary and her friends, but the most wonderful thing about him, the outstanding touch, was that he liked me. I still cannot understand why unless it was that he was overcome by so much untrammeled girlishness. He took me to dinner, dancing and the movies and I fell head over heels in love, to his evident delight, and when I was eighteen we were married. Bob was thirteen years older than I but a far cry from the Smith Brothers.

Why do more or less intelligent people go on honeymoons, anyway? I have yet to find a couple who enjoyed theirs. And, if you have to go on a honeymoon, why pick quaint, old-world towns like Victoria, B. C., which should be visited only with congenial husbands of at least one year's vintage or relatives searching for antiques.

We honeymooned in Victoria for a week and though I had visited there many times previously, I was surprised that I hadn't noticed what a dull place it was. Nothing to do. Victoria's idea of feverish gaiety is Thé Dansant at a hotel where

Canadian women in white strap slippers, mustard-colored suits and berets, dip and swirl with conservative Canadian men. We spent one afternoon at Thé Dansant but there was a noticeable lack of hilarity at our table. Bob, that dear, gay, understanding companion of our courtship days, sat with chin on chest staring moodily at the dancers while I ate. I ate all of the time we were in Victoria. I was too fat and I wanted desperately not to eat and be willowy and romantic but there seemed nothing else to do. Bob ate almost nothing and looked furtive like a trapped animal. I guess it is quite a wrench for a bachelor to give up his freedom, particularly when, every time he looks at his wife, he realizes that he is facing a future teeming with large grocery bills.

On the boat going up to Victoria, Bob seemed to be well established in the insurance business and held forth at some length on premiums, renewals and "age 65," and I determined to ask Mother just how much I should learn about insurance in order to be helpful but not meddlesome, and wondered what the wives of insurance men were like for friends. On the way back from Victoria, Bob talked of his childhood on a wheat ranch in Montana, his days at agricultural college and his first job as supervisor for a large chicken ranch. When he spoke of the wheat ranch it was with about as much enthusiasm as one would use reminiscing of the first fifteen years in a sweat shop and I gathered that he thought farming hard, thankless work. But then he began on the chicken ranch job, sorting over the little details with the loving care usually associated with first baby shoes. When he reached the figures—the cost per hen per egg, the cost per dozen eggs, the relative merits of outdoor runs, the square footage required per hen—he recalled them with so much nostalgia that listening to him impartially was like trying to swim at the edge of a whirlpool. He told me at last that he had found a little place on the coast, where he often went on business, that was ideally situated for chicken ranching and could be bought for almost nothing. "What did I think about it?" What did I think about it? Why, Mother had taught me that a husband must be happy in his work and if Bob wanted to be happy in the chicken business I didn't care. I knew how to make mayonnaise and mitre sheet corners and light candles for dinner, so, chickens or insurance, I could hold up my end. That's what I thought. That's what a lot of women think when their husbands become dewy-eyed at the sight of their breakfast eggs and start making plans for taking the life savings and plunging into the chicken business.

Why in God's name does everyone want to go into the chicken business? Why has it become the common man's Holy Grail? Is it because most men's lives are shadowed by the fear of being fired—of not having enough money to buy food and shelter for their loved ones and the chicken business seems haloed with permanency? Or is it that chicken farming with each man his own boss offers relief from the employer-employee problems which harry so many people? There is one thing about the chicken business: if a hen is lazy or uncooperative or disagreeable you can chop off her head and relieve the situation once and for all. "If that's the way you feel, then take that!" you say, severing her head with one neat blow. In a way I suppose that one factor alone should be justification enough for most men's longing for chickens, but again I repeat, why chickens? Why not narcissus bulbs, cabbage seed, greenhouses, rabbits, pigs, goats? All can be raised in the country by one man and present but half the risk of chickens.

The next morning after our return to Seattle, the alarm went off with a clang at six-thirty; at six-thirty-one Bob, clad in a large wool plaid shirt, was stamping around the kitchen of our tiny apartment making coffee, and demanding that I hurry. At eight-forty-five we had driven twelve miles and were boarding a ferry as the first lap in our journey to see the "little place."

It was one of our better March days—it was, in fact, one of the March days we have up here which deceives people into thinking, "With spring like this we are sure to have a long, hot summer," and into stocking up on halters and shorts and sunglasses. Then later, summer appears wan and shaking with ague and more like February. This March day, though, was strong and bright and Bob and I spent the long ferry ride walking the decks and admiring the deep blue waters of Puget Sound, the cerulean sky, densely wooded dark-green islands which floated serenely here and there, and the great range of Olympic Mountains obligingly visible in all of their snowy magnificence. These Olympics have none of the soft curves and girlish plumpness of Eastern mountains. They are goddesses, full-breasted, broad-hipped, towering and untouchable. They are also complacent in the knowledge that they look just as mountains should.

We were the only passengers on the large, crowded ferry who took a breath of fresh air or even glanced at the spectacular scenery. The rest of them, business men, salesmen, farmers' wives, mill workers and Indians, either remained below in their cars or the bus which boarded the ferry or huddled in the hot lounges and read newspapers in a bad light. They were a forbidding-looking bunch and Bob and I ran a gantlet of ferociously hateful looks when we came heartily inside, after half an hour or so, stamping our cold feet and slamming the doors and searching hopefully for coffee. We found the coffee, dark green and lukewarm, in the galley and drank it to the morose accompaniment of two farmers' wives discussing "the dreen tubes in Alice's incision." Bob had been smoking when we came in and apparently no one noticed it but when, halfway through my cup of coffee, I lit a cigarette, one of the farmers' wives snatched off her manure-colored hat and began fanning the air violently in my direction, meanwhile uttering little hacking coughs. I continued to smoke, so the other woman picked up a newspaper and waved it so vigorously that I was afraid she'd sweep our coffee cups into our laps. Bob hissed at me, "Better put out your cigarette," and I hissed back, "I wish I had a big black cigar," and he looked at me reproachfully and led me outside and handed me a small pamphlet which I thought might be a religious tract but turned out to be a small travel booklet describing the country, in the depths of which the prospective ranch was hidden. It was a brochure of superlatives. "The Olympic Mountains are the most rugged mountains on the North American Continent . . . the largest stand of Douglas fir in the world . . . three million acres, two and a half million of which are wild . . . Cape Flattery is the most westerly point in the United States . . . the greatest fishing fleets on the Pacific fish from Cape Flattery." The little book stated that here was nature at her most majestic, that opportunity was pounding at the door, natural resources were pleading to be used and scenic drives aching to be driven. I thought the whole thing slightly hysterical but then I hadn't seen the country. Now I know that that country is describable only by superlatives. Most rugged, most westerly, greatest, deepest, largest, wildest, gamiest, richest, most fertile, loneliest, most desolate—they all belong to the coast country.

The ferry landed, we drove ashore and made a circuit of the two streets which comprised Docktown. There were a great sawmill, a charming old Victorian hotel with beautifully cared for lawns and shrubs, a company store, a string of ugly company houses, and a long pier where freighters were being loaded with lumber by an alarmingly undecided crane that paused first here, then there and finally dumped a gigantic load of planks almost on top of the longshoremen. Curses flew up like sparks from the men as they scattered to safety but in a moment or so the air cleared and they were back at work. Cranes and piledrivers can keep me at a pitch of nervous excitement for hours and hours and when I finally do tear myself away it is always with the conviction that the operator is going to find the operation very difficult without my personal supervision. I would have been content to lean on the sun-warmed railing of the ferry dock, smelling that delicious mixture of creosote, cedar and seaweed which characterizes coast mill towns, and watching the cranes for the rest of the day; but Bob warned me that we had a long drive ahead of us and if we intended to return that night we should get started.

The road out of Docktown was dangerously curved and not too wide and alive with cars, trucks and logging trucks with terrific loads and terrible trailer tails that switched and slithered behind them. Everyone drove as if he were going to a fire and on the wrong side of the road, and we were warned of approaching corners by the anguished screams of tires and brakes. Bob is an excellent driver but he was hard put to it to hold his own when a logging truck carrying three of the largest logs of the largest stand of Douglas fir in the world came winging around a curve and we had to leap the bank and scurry for the woods to avoid being smashed into oblivion by the playful trailer. The driver leaned out and grinned and waved at us and then went careening off down the road. We backed carefully onto the road again and trundled sedately off, hugging the bank nervously when we spotted another logging truck. After a while we left the woods and began skirting a great valley where emerald winter wheat, the velvety blackness of plowed fields and the tender green of new pastures checkered the bottom land. This was a dairy country and the smallest farms ran to three hundred and fifty acres. The houses, for the most part unattractive boxlike abodes, close to the road and unadorned with flowers or shubbery, were across the road from their farm lands, their back porches snuggled against the blue-black tree-covered hillsides. The barns, silos, bunkhouses and outbuildings, magnificent structures of generous proportions were on the valley side. I thought this arrangement had something to do with keeping the cattle out of the house until Bob informed me that the road had been put in after the ranches were laid out.

Black and white Holstein cattle and deserted farms seemed to dominate the landscape and one was responsible for the other, according to Bob. This valley once boasted some of the finest Holstein herds in the country and the farmers invested heavily in breeding stock, but when the Holstein market collapsed some years back, many of them went bankrupt. The farmers the Holstein market didn't get were soon put in their places by contagious abortion and tuberculosis in their herds and a Government drainage ditch, the assessments on which were terrific, on their lands. In addition to this they had the ever-present problem of marketing and were either at the mercy of the local creameries and cheese factory or occasional city firms, none of which, the farmers said, gave them a square deal. Bob did not waste much sympathy on them, however; he said they were hopelessly unprogressive and many were using biblical methods of production and complaining because they couldn't compete in up-to-date markets.

I had noticed wisps of smoke rising from the ground in the farthest fields. "Burning peat," Bob explained. "One of the great tragedies of this country. Years ago some of the farmers, in an effort to clear the practically unclearable peat land, set fire to some of the huge piles of logs, roots and trees unearthed during plowing. When the roots and stumps had burned the farmers were surprised to find that the land itself was burning and that ditching, plowing and wet sacks were ineffective agents in putting it out. After much experimenting they learned that by digging four-foot-deep drainage ditches around a small area at a time, they could control the fire but this was such an undertaking that in most cases they let 'er burn."

"Isn't the land arable after the fires have burned out?" I asked.

"Unfortunately, not for years and years because peat burns deep down to a light feathery ash which will not bear the weight of a horse or a tractor. Hand-cultivated, it will grow potatoes almost as big as watermelons and about as watery, too," Bob concluded dismally.

"Look at those fields," I exclaimed pointing to plowed fields as black as licorice. "That soil must be terribly rich."

"It's rich all right," Bob said, "but it's peat land and hellishly expensive to clear and drain. You clear and plant a field and the next year your plow digs up a stump every three feet and you have to clear all over again. Every acre of it has to be tiled drained, too."

After that, for a time we drove along in silence while the unconquerable peat lay black and scornful in the valleys and the unconquerable forests thundered down at us from the hills.

"This land resents civilization and it isn't a little futile stick-out-the-tongue kind of resentment, but a great big smashing resentment that is backed by all the forces of nature," I thought, huddling down into my coat and hoping we'd soon come to a town.

We did, and it boasted the mad confusion of four enterprises—a hotel, a barbershop, a gas station and a country store and post office. In addition there were a dear little graveyard and an imposing brick schoolhouse. Five roads led away from this small town but Bob didn't hesitate. He chose one pointing southwest toward the frosty Olympic Mountains. For the next several hours we saw no more towns, only crossroads stores; rich valleys separated by heavily wooded hills; herds of cattle and widely spaced farm houses. We had nosed our way into the foothills of the Olympics while we were still in the farming country and it wasn't until I looked from the car window and saw, far below the road, a frustrated little mountain stream banging its head against immense canyon walls that I realized that we were in the mountains proper. Yellow highway signs announcing WINDING ROAD appeared at intervals and Bob put the car in second and then low gear as we spiraled forward and upward. We were climbing but seemed to be getting nowhere for we were walled in on all sides by the robust green mountainsides and only by sticking my head clear out of the window was I able to peer up and see the sky. Two or three hundred million board feet of Douglas fir later, we turned off the main highway onto a dirt road and jounced and skidded our way at last to the "little place."

On first sight it looked distressingly forlorn, huddled there in the laps of the great Olympics, the buildings grayed with weather, the orchard overgrown with second-growth firs, the fences collapsing, the windows gaping. It was the little old deserted farm that people point at from car windows, saying, "Look at that picturesque old place!" then quickly drive by toward something not so picturesque, but warmer and nearer to civilization. Bob halted the car to take down the rails of the gate and I looked morosely around at the mountains so imminent they gave me a feeling of someone reading over my shoulder, and at the terrific virility of the forests, and I thought, "Good heavens, those mountains could flick us off this place like a fly off their skirts, rearrange their trees a little and no one would ever be the wiser." It was not a comforting thought and the driveway, which proved to be a rather inadequate tunnel under the linked arms of two rows of giant trees, did nothing to dispel it. Heavy green branches lashed the top of the car and smaller twigs clawed at the windows and the car wheels churned and complained on the slick dry needles. We drove for perhaps a quarter of a mile like this and then abruptly the trees stopped and we were in the dooryard of the farm, where a great-grandfather of a cherry tree, hoary with bloom, stood guard over the huddled buildings.

I'm not sure whether it was the cherry tree or the purple carpet of sweet violets flanking the funny silvery woodshed, or the fact that the place was so clean, not a scrap of rubbish, not a single tin can, but it suddenly lost its sinister deserted look and began to appear lonely but eager to make friends. A responsive little farm that with a few kindnesses in the way of windows and paint and clearing might soon be licking our hands.

While I stood in the dooryard "feeling" the place, Bob was bounding around with a hammer, pounding the walls and calling happily, "Look, Betty, hand-hewn-out-of-cedar logs, and sound as a nut." The hand-hewn cedar shakes which covered the sides and roof had worked loose in several places and Bob pulled them off to show me the cedar logs and the axe marks.

The house, evidently begun as a log cabin about twenty feet by twenty and added on to at either end, was beautifully situated on a small rise of ground from which an old orchard, peering out from the second-growth fir, sloped gently down to a small lake or large pond. The original cabin was the living room with windows on the north and south sides and a thin rickety porch across the front. It faced south, across the orchard, to the pond and of course the mountains. The mountains were everywhere—I'd start to turn around, come up against something large and solid and wham! there was a mountain icily ignoring me.

Opening off the living room on the right, with windows north, west and south, we found a bedroom with roses and honeysuckle vines in heaps on the floor below the windows, as though they had climbed up to peek in and had fallen over the sills. Down three steps and to the left of the living room were an enormous square kitchen with windows east and north and a pantry the size of our apartment in town, with three windows facing east. Jutting off the kitchen toward the front was a bedroom with windows looking east and south. Up a creepy flight of stairs from the living room were two tiny slope-ceilinged bedrooms. Under the front porch we discovered a bat-hung cellar, and to one side of the kitchen, forming an ell with the living room, an entryway and wood room.

A very large, very surly and slightly rusty range was backed defiantly against the north wall of the kitchen—otherwise the place was empty. The floors were warped and splintery—the walls were covered with carefully tacked newspapers dated 1885.

At first glance the outbuildings seemed frail and useless, but closer examination revealed fine bone structure in the way of uprights, beams and stringers and so we were able to include in the assets of the place, a very large barn, two small chicken houses, a woodshed and an outhouse. The assets also included ten acres of land showing evidences of having once been cleared, and thirty acres of virgin timber, cedar, fir and hemlock—some of it seven feet and more in diameter. Scattered over the ten cleared acres, like figures in a tableau, were the dearest, fattest, mostly perfectly shaped Christmas trees I have ever seen. Each one was round and full at the bottom and exquisitely trimmed with brown cones. I was caressing and exclaiming over these when Bob told me that such little jewels of trees are cut by the hundreds of thousands by Christmas tree dealers, who pay the farmers two cents each for them. Incredible that anyone who professed a love of the soil would sanction such vandalism and for such a paltry fee.

At the edge of the clearing and sheltered by one of the great black firs, we found an old well. It was half full of water, but the intake was a tiny trickle instead of a robust gush which this season warranted, so Bob decided it had been abandoned and we looked elsewhere for water. We found a larger, more substantial spring at the foot of the orchard, feeding the lake, but as it had not been boxed in and showed no other signs of use, either it was a thing of recent origin or suffered from summer complaint—time would tell. It did too, and water became one of the major obsessions of my life.

We threaded our way through the orchard and found slender fruit trees bravely blossoming with frail hands pushing futilely against the dark green hairy chests of the invading firs. The firs were everywhere, big and virile, with their strong roots pulling all of the vitality out of the soil and leaving the poor little fruit trees only enough food and light to keep an occasional branch alive. These were no kin to the neatly spaced little Christmas tree ladies of the back pasture. These were fierce invaders. Pillagers and rapers.

The more we walked around, the stronger became my feeling that we should hurry and move in so that we could help this little farm in its fight against the wilderness. Bob was overjoyed when I told him of this feeling and so we decided to buy it at once.

For the forty acres, the six-room log house, the barn, two

small chicken houses, woodshed, outhouse and the sulky stove, the mortgage company was asking four hundred and fifty dollars. Between us and by pooling all savings accounts, wedding presents, birthday presents and by drawing on a small legacy which I was to get when I became twenty-one we had fifteen hundred dollars. We sat in the sunny doorway under the cherry tree, used a blue carpenter's pencil and shingle and decided that we would pay cash for the farm; put seven hundred dollars in the bank to be used to buy, feed and raise three hundred and fifty pullets; and we would use the rest to fix up the buildings. Fuel and water were free and we'd have a large vegetable garden, a pig to eat leavings, a few chickens for immediate eggs and Bob could work occasionally in one of the sawmills to eke out until the chickens started to lay. Written out in blue pencil on the weathered shingle it was the simplest, most delightful design for living ever devised for two people.

We left then and hurried home to put our plans into action. The next morning Bob paid the $450, and brought home the deed. The following week we borrowed a truck, loaded on everything we possessed and left for the mountains to dive headfirst into the chicken business.

"Which all goes to show," I said, "that preparing a girl for marriage before she marries is *battre l'eau avec un bâton";* and as Bob often remarked later, "To prepare a girl for marriage after she is married is *vouloir rompre l'anguille au genou."*

3

"Who, Me?" or *"Look 'Peasant,' Please!"*

"WHO, ME?" I asked when we were moving and Bob pointed casually to a large chest of drawers and said, "Carry that into the bedroom."

"Who else?" he snapped and my lower lip began to tremble because I knew now that I was just a wife.

"Who, me?" I asked incredulously as he handed me the reins of an enormous horse which he had borrowed from a neighbor, and told me to drive it and a heaving sled of bark to the woodshed while he gathered up another load.

"Yes, you!" he roared. "And hurry!"

"Not me!" I screamed as he told me to put the chokers on the fir trees and to shout directions for the pulling as he drove the team when we cleared out the orchard. "Yes, you! I'm sure you're not competent but you're the best help I can get at present," and Bob laughed callously.

"Hand me that hammer. Run into the house and get those nails. Help me peel this stringer. Hurry with those shakes. Put your weight on this crowbar. Stain that floor while I lay this one. You don't measure windows that way, bonehead. Help me unload this chicken feed. Run down and get a couple of buckets of water."

"If I can handle the plow, surely you might manage the horse more intelligently!"

"Go get those seeds. It's time to fill the baby chicks' water jugs. Bring me some of those two-by-fours. Cut me about twenty-five more shakes. Don't be such a baby, bring them up *here.* I'm not climbing down from this roof everytime I want a nail."

And that's the way it went that first spring and summer. I alternated between delirious happiness and black despair. I was willing but pitifully unskilled. "If only I had studied carpentry or mule skinning instead of ballet," I wailed as I teetered on the ridgepole of the chicken house pounding my already mashed thumbs and expecting momentarily to swallow the mouthful of shingle nails which pierced my gums and jabbed into my cheeks.

"You're coming along splendidly," said Bob kindly and he could afford to be kind for his work was like the swathe of a shining sharp scythe. He was quick, neat, well-ordered and thorough. My efforts were more like shrapnel—nicest where they didn't hit. Bob pounded nails with a very few, swift, sure strokes, right smack on the head. I always tried to force my nails in sideways and my best efforts look hand adzed. Bob sawed lightly, quickly and on the line. Zzzzzzz—snap and the board was through with the sawdust in orderly little heaps on either side. My saw rippled in, was dragged out, squealed back and when I got through Bob said, "How in God's name did you get that scallop in there?" He had the temperament and the experience and all I had was lots of energy.

The first day we moved all of the furniture into the house and I thought that the next day we would start putting in windows, laying new floors, painting woodwork and sheathing the walls. That's what I thought. The next day we started building a brooder house because to get started with the baby chickens was the important thing. Bob cut the log stringers about twenty-five feet from the building site, hauled the rest of the lumber from the mill and I split the shakes with a dull chipped frow and many vigorous curses. We built the brooder house in the prettiest part of the orchard, facing the pond and the mountains, and its newness was so incompatible with the other silvery buildings that I suggested to Bob that we plant a few quick growing vines and perhaps a shrub or two to tone it down a little. He was as horrified as though I had suggested bringing potted plants into a surgery. "Brooder houses are built on skids so that they can be moved from place to place as baby chickens must have new UNTAINTED soil," he said. This still seems an unnecessary precaution to me, for the land up there was all of it so untainted, so virginal, that I expected the earth to yell "Ouch" when we stuck a spade into it and any germ that could have survived the rigors of that life would have been so big and strapping we could have seen it for blocks. However, the brooder house was built on runners and remained an eyesore, except for the shake roof, which weathered beautifully.

When the brooder house was finished and the seven hundred and fifty yeeping chicks and the brooder were installed therein, I thought we would then begin on the house. The nights were very cold and it rained at least three of the seven days a week and I thought we might pamper ourselves with a few windows and doors. That's what I thought. The most important thing was to build two small pullet houses and to whitewash the walls of and lay a new floor in one of the small chicken houses so that the cockerels would be comfortable while fattening. We built the cockerels a nice yard also and then we remodeled the other small chicken house for the baby pig because the pig must be comfortable and protected from the cold night air and the damp day air. By the time we finished those buildings it was May. A cold damp May with so much rain that mildew formed on our clothes in the closets and the bedclothes were so clammy it was like pulling seaweed over us.

"Now," I thought, "we have all the livestock warm and comfortable, surely it is at last time to fix the house." That's what I thought. It was time to plow and plant the garden. I had read that the rigors of a combination of farm and mountain life were supposed eventually to harden you to a state of fitness. By the end of those first two months, I still ached like a tooth and the only thing that had hardened on the ranch was Bob's heart.

Right after breakfast one May morning he drove into the yard astride a horse large enough to have been sired by an elephant. Carelessly looping the reins over a gatepost he informed me that I was to steer this monster while he ran along behind holding the plow. All went reasonably well until Birdie, the horse, stepped on my foot. "She's on my foot," I said mildly to Bob who was complaining because we had stopped. "Get her off and let's get going," shouted the man who had promised to cherish me. Meanwhile my erstwhile foot was being driven like a stake into the soft earth and Birdie stared moodily over the landscape. I beat on the back of her knee, I screamed at her, I screamed at Bob and at last Birdie absentmindedly took a step and lifted the foot. I hobbled to the house and soaked my foot and brooded about men and animals.

That evening Bob and I sat opposite each other for two hours sorting and cutting seed potatoes. The billing and cooing of the newly married love birds consisted of, "This is an eye. An eye is a sprout. A sprout makes a plant. Each piece must have three eyes." "Is this an eye?" "No!" "Why not?" "Oh, God!"

I thought, as I lay in bed that night listening to Bob snore and the coyotes howl, "The lives of Elizabeth Browning and Beth in *Little Women* weren't half bad. I wonder if Elizabeth would have been so gentle and sweet if in answer to one of her Bob's whims she had had her foot stepped on by a horse."

When the garden, about 50 feet by 350 feet, had been plowed, disked, harrowed and dragged until the deep brown loam was as smooth as velvet, it was planted to peas, beets, beans, corn, Swiss chard, lettuce, cabbage, onions, turnips, celery, cucumbers, tomatoes and squash. The preparing process was repeated on an acre or so in the back field which was planted to potatoes, kale, mangels and rutabagas.

Then I was drafted into the stump-pulling department. The scene was the orchard and my part in the activity was to try and grab the chain as the horse walked by, fasten it around the trunk of a fir tree before the horse shifted her position and it wouldn't reach, shout "Go ahead" to Bob and forget to get out of the way of the heavy sprays of damp loam. Clearing land is very satisfying work because you have something definite to show for your efforts, even if it is only a large hole. In the orchard it was wonderful to watch a little fruit tree huddle fearfully as we worked to remove a large bullying fir; then when with a last grunt the protesting fir was dragged away and the earth patted back on the fruit trees' roots, to watch the little tree timidly straighten up, square its shoulders and stretch its scrawny limbs to the sun and sky.

When the last fir had been hauled away to the stump pile the other side of the east fence to be burned the next winter, Bob and I pruned away dead limbs on the fruit trees and made guesses as to varieties of fruit. We knew by the blossoms that there were early and late apples, cherries, pears, plums and prunes, but we had no way of knowing which trees would bear and what they would bear. Unfortunately the sturdiest trees turned out to be the poorest varieties and many of the trees bore nothing or merely two or three wizened nubbins. By fall, however, we were sure that we had two Gravenstein, one Wealthy, one Baldwin, one Winter Banana and two Yellow Transparent apple trees; two Italian prune trees and a green-gage plum tree; three or four Bartlett and two Seckel pear trees, several Bing cherry trees which bore a cherry per tree, and the grandfather cherry tree in the backyard that was some obscure variety which no visiting nurseryman was ever able to identify. This tree was a prolific bearer, a late ripener (around the last of August) and the cherries were large, bright red and crunchy with juice and sugar. Because of its great vigor and evident health, it was the only tree on the ranch which escaped the ravages of Bob's pruning. While we were still in the process of clearing the orchard, Bob saw an advertisement in a farm magazine for a pruning guide and it seemed a sensible thing to have, so he sent for it. It came promptly and with it also came a large colored sheet of required tools and implements. Bob sent for them all. Curved saws, short shears, long shears, short clippers, long clippers, knives and machetes. All implements of destruction. That first year Bob practised on the vines and shrubs, reducing them to stumpage with a few sure strokes. The next spring he fixed the orchard. "This must come off," he would say squinting first at the pruning guide and then at the lone live limb on the tree. And off it came and the tree died.

"Suckers!" he sneered, grasping the only live shoots on a quivering plum tree, and snipping them off with his vicious clippers. The tree died. Bob was hurt. "That's what they said to do," and he pointed out the instructions. "Yes, but perhaps they refer to new, young trees," I suggested. "Our trees are probably forty or fifty years old and weak and starved."

"And better off dead," Bob concluded, but seemed relieved when I showed him that the climbing rose on the kitchen window was going to live after all.

When the gardens were finished and the orchard had been cleared and plowed, we started on the big chicken house. Up to this time we had been buying all of our building materials from the Docktown sawmill and our groceries from the company store, but now we had to have special items like glass cloth and heavy mesh wire so we took a trip to "town."

"Town" was the local Saturday Mecca. A barren old maid of a place, aged and weathered by all the prevailing winds and shunned by prosperity. Years ago the Town with her rich dot of timber and her beautiful harbor was voted Miss Pacific Northwest of 1892 and became betrothed to a large railroad. Her happy founders immediately got busy and whipped up a trousseau of three- and four-story brick buildings, a huge and elaborate red stone courthouse, and sites and plans for enough industries to start her on a brilliant career.

Meanwhile all her inhabitants were industriously tatting themselves up large, befurbelowed Victorian houses in honor of the approaching wedding. Unfortunately almost on the eve of the ceremony the Town in one of her frequent fits of temper lashed her harbor to a froth, tossed a passing freighter up onto her main thorofare and planted seeds of doubt in the mind of her fiancé. Further investigation revealed that, in addition to her treacherous temper, she was raked by winds day and night, year in and year out, and had little available water. In the ensuing panic of 1893, her railroad lover dropped her like a hot potato and within a year or so was paying serious court to several more promising coast towns.

Poor little Town never recovered from the blow. She pulled down her blinds, pulled up her welcome mat and gave herself over to sorrow. Her main street became a dreary thing of empty buildings, pocked by falling bricks and tenanted only by rats and the wind. Her downtown street ends, instead of flourishing waterfront industries, gave birth to exquisite little swamps which changed from chartreuse to crimson to hazy purple with the seasons. Her hills, shorn of their youthful timber in preparation for a thriving residential district, lost their bloom and grew a covering of short crunchy grass which was always dry and always yellow—lemon in spring, golden in summer and fall. She wore her massive courthouse like an enormous brooch on a delicate bosom and the faded and peeling wedding houses grew clumsy and heavy with shrubbery and disappointment.

Of late years a small but gay army post and a thriving branch of the Coast Guard settled within arm's length, but the Town shunned their advances, preferring just to keep body and soul together through her little ordinary businesses. These were succored by the surrounding country and the mountain dwellers who referred to her disparagingly as the world's only lighted cemetery. All except me, that is, and to me "Town" spelled L-I-F-E!

I loved the long sweeping hill that curved down to Town: I loved the purply-green marshes we crossed at the bottom of the hill; I loved the whitecaps in the harbor and the spray lacing the edges of the streets; I loved the buildings squatting along the water on the main thorofare, their faces dirty but earnest, their behinds spanked by icy waves; and I loved the Town's studied pace, her unruffled calm, her acceptance of defeat.

We drove around her quiet streets, over her lovely hills and looked at her wonderful views, then we went into her shops, and we knew instantly that her customers were farmers and Indians. The grocery store smelled like sweat, cheese, bakery cookies and manure. The drug store smelled like licorice, disinfectant, sweat and manure. The hardware store smelled like commercial fertilizer, sweat and manure. The only place which managed to rear its head above the customers was the small candy store catering mostly to townspeople. I breathed in great draughts of its rich fudgy smells and bought a bag of vanilla caramels (pronounced "kormuls" by the proprietress) which had evidently never hardened for they stuck to the brown paper bag and to each other so stubbornly that I had either to turn back the bag and lick the brown lumpy mass therein or to pry off little bits of candy and brown paper and eat both. Finally I threw it away en masse and watched the sea gulls scream and swoop for it and was disappointed that they didn't fly up again all stuck together by the beaks.

We had three strokes of luck in Town on that visit. One was that a check of our bank balance found that it tallied with our check stubs and was holding up surprisingly well. Another was a tip on where to buy a dozen laying pullets for ten dollars. And the last was a ranch-warming present of a gallon of moonshine from the best of the local moonshiners, of which there seemed to be hundreds. The moonshine in a gallon jug was a dark amber color and had a hot explosive smell. We had a drink before dinner that night and it went down with lights flashing like marbles in a pinball game.

The next morning we had breakfast in that filmy period just before daylight and were busily at work on the big chicken house before sunrise. The work was hard and the task large but Bob and I, or rather, Bob impeded by me, had first to remove all of the big barn's viscera; then put in nests, dropping boards, roosts and windows on three and one-half sides (the other half was the doorway); and to install Healtho-Glass in the windows. Healtho-Glass was a glass cloth which, so the advertisements said, sorted out the sunlight and let in *only* the health-giving *violet rays.* The first day we put it up I half expected to find the chicken house suffused in soft lavender light and the hens scratching around under a purple spot. Actually Healtho-Glass gave the same type and amount of light as frosted lavatory windows.

Down the middle of the old barn were log uprights. We naturally didn't jerk these out although I thought it a fine idea until Bob pointed out drily that they held up the roof. We built mash hoppers between the uprights, whitewashed the walls even unto the rafters, swept and scraped the hard dirt floor—the barn, like most things there in the mountains, had gone barefoot all of its life and the soles of its feet were as tough and smooth as leather—and it turned out to be a very useful, though unorthodox, chicken house where we kept as many as fifteen hundred hens.

The first day the chicken house was finished Bob drove to Town and bought the twelve Rhode Island Red pullets for ten dollars and we immediately turned them into the great new house, where they rattled around like beads in an empty bureau drawer and, being chickens, instead of laying in the row upon row of convenient new nests, they laid their eggs on the dropping boards at the entrance of rat holes or out in the yard.

It was late summer before we even started on our house. We laid new floors; put in windows; kalsomined the walls; fixed broken sills and sagging doors; put in a sink (without water but with a drain) and made other general repairs and, though it looked about as stylish as long underwear in its gray sturdiness, it began to feel like home. The kitchen had two armchairs and a rocking chair, a big square table, rag rugs and the stove. The kitchen was the hub of all our activities. We kept the egg records there—we wrote our checks—made out our mail orders—read our mail—ate—washed—took baths—entertained—planned the future and discussed the past. We began the day in it at 4 A.M. and we ended it there about eight-thirty by shutting the damper in the stove just before blowing out the lamp. The rest of the house was clean and comfortable and unimportant.

We traded our car for a Ford pick-up truck. We traded a valley farmer a waffle iron and toaster (wedding presents) for a dragsaw. We traded electric lamps (wedding presents) for gasoline lanterns, kerosene lamps and sad irons. We bought tin washtubs and a pressure cooker.

We hewed a road into the virgin timber at the back of the ranch and drove the truck out there, loaded with axes, mauls, wedges, peaveys, oil and gasoline and we sawed shake bolts from fallen cedars four feet in diameter and straight grained. We sawed fallen firs, six and seven feet in diameter and conky in the middle, for wood. The dragsaw barked and smoked dangerously but its sturdy little arm pulled the blade back and forth with speed and skill and the great wooden wheels rolled off and Bob split them with the sledge hammer and the wedge, and I stacked them in the truck and gathered bark.

The woods were deep and cool and fragrant and treacherous with underbrush, sudden swamps and roots. Laden with a six-inch-thick hunk of bark and a wedge of wood I would start toward the truck, step on what I took for a hummock, go knee-deep in water, get slapped in the face by the Oregon grape and salal bushes, and peel the skin off my forearms as I fell with the wood. The next two or three trips would be without mishap, then just as I reached the truck, overconfident and overloaded, I'd catch the toe of my shoe in a root and fall flat. I learned the inadequacy of "Oh, dear!" and "My goodness!" and the full self-satisfying savor of sonofabitch and bastard rolled around on the tongue. I also learned the meaning of a great many homely phrases that first spring and summer. Things like "Shoulder to the wheel," which meant actually my shoulder to the wheel of the truck while Bob raced the motor and tried to pull it out of a hole. "Two honest hands" which were Bob's and mine, hoeing, weeding, chopping, feeding, caring for and cleaning. "Teamwork" was Bob and Birdie and me pulling stumps. "Woman's work is never done" signified the dinner dishes which I washed and dried while Bob smoked his pipe and took his ease.

I believe that long-suffering Bob learned best the meaning of the one about a wife being an impediment to great enterprise.

There were many nights when I was so tired I couldn't sleep and I tossed fitfully, and hurt in numerous places and thought, "And they call this living?" The next morning I would get up sore and stiff and grumpy and then suddenly the windows in the kitchen would begin to lighten a little and I knew it was time for the sunrise. I'd rush outdoors just as the first little rivulets of pale pink began creeping shyly over the mountains. These became bolder and brighter until the colors were leaping and cascading down the mountains and pouring into the pond at the foot of the orchard. Faster and faster they came until there was a terrific explosion of color and the sun stood on the top of the mountains laughing at us. The mountains, embarrassed at having been caught in their nightdresses rosy with sleep, would settle back with more than their accustomed hauteur, profiles cold and white against the blue horizon. Then from the kitchen would come the smell of coffee, that wonderful heartwarming smell, and I'd think "Life is wonderful!" as Bob came whistling in to breakfast.

By fall our potatoes were dug, our pullets were laying, our roosters had been fattened and sold and we were really chicken farmers keeping detailed egg records and netting around $25.00 a week from our three hundred and fifty hens. Several thousand sets of new muscles had stopped aching, the blisters on my hands were healing and one night I lay in bed beside Bob and watched a full moon come up from behind the black hills—there would be a frost before morning—listened to Bob's breathing, so deep and peaceful—heard the stove make occasional crunching noises as it ate into its night-load of bark—overheard a little mouse scratching gently and thought, "This is the life, after all."

And then winter settled down and I realized that defeat, like morale, is a lot of little things.

PART TWO

November

No sun—no moon—no morn—no noon,
No dawn—no dusk—no proper time of day,
No warmth—no cheerfulness—no healthful ease,
No road, no street, no t'other side the way,
No comfortable feel in any member—
No shade, no shine, no butterflies, no bees,
No fruits, no flowers, no leaves, no birds, November!

——HOOD

4

The Vanquished

DESPITE its location, I never had the feeling that our small ranch was nestled on the protective lap of the Olympic Mountains. There was nothing protective about them. Each time I looked out of a window or stepped out of doors, I was confronted by great, white, haughty peaks staring just above my head and doing their chilly best to make me realize that that was once a very grand neighborhood and it was curdling their blood to have to accept "trade." We were there with our ugly little buildings and livestock, but, by God, they didn't have to associate with us or make us welcome. They, no doubt, would have given half their timber if they could have changed the locale to Switzerland and brushed us off with a nice big avalanche.

All that first spring and summer they were obviously hostile but passive. With the coming of September they pulled mists down over their heads like Ku-Klux hoods and began giving us the old water cure.

It rained and rained and rained and rained. It drizzled—misted—drooled—spat—poured—and just plain rained. Some mornings were black and wild, with a storm raging in and out and around the mountains. Rain was driven under the doors and down the chimney, and Bob went to the chicken house swathed in oilskins like a Newfoundland fisherman and I huddled by the stove and brooded about inside toilets. Other days were just gray and low hanging with a continual pit-pat-pit-pat-pitta-patta-pitta-patta which became as vexing as listening to baby talk. Along about November I began to forget when it hadn't been raining and became as one with all the characters in all of the novels about rainy seasons, who rush around banging their heads against the walls, drinking water glasses of straight whiskey and moaning. "The rain! The rain! My God, the rain!"

In case you are wondering why I didn't take a good book, settle down by the stove and shut-up, I would like to explain that Stove, as we called him, had none of the warm, friendly qualities ordinarily associated with the name. In the first place he was too old and, like some terrible old man, he had a big strong frame, a lusty appetite and no spirit of cooperation. All attempts to get Stove to crackle and glow were as futile as trying to get the Rock of Gibraltar to giggle and cavort. I split pure pitch as fine as horsehair and stuffed his ponderous belly full, but there was no sound and no heat. Yet, when I took off the lids the kindling had burned and only a few warm ashes remained. It was as mysterious as the girl in high school who ate enormous lunches without apparently chewing or swallowing.

Incongruously, things did boil on Stove. This always came as a delightful shock, albeit I finally stopped rushing to the back door and shouting hysterically to Bob, quietly and competently at work, "The water is BOILING!" as I had done for the first few hundred times I had witnessed this miracle.

I put my first cake into the oven with such a sense of finality that I almost added a Rest-in-Peace wreath, and I felt like Sarah Crewe when I came in from the chicken house and the air was vibrant with the warm spicy smell of baking.

On the coldest dreariest mornings Stove sulked all over his end of the kitchen. He smoked and choked and gagged. He ate load after load of my precious live bark and by noon I could have sat cross-legged on him and read *Pilgrim's Progress* from cover to cover in perfect comfort.

Stove was actually a sinister presence and he was tricky. The day we first looked at the place, I remarked that he seemed rather defiantly backed up against the wall, but such an attitude could come from neglect, I thought, and so when we moved in the first thing I did was to clean his suit, take all the rust off his coat and vest, blacken every inch of him, except his nickel which I polished brightly, and then I built my first fire, which promptly went out. I built that fire five times and then Bob came in and poured about a gallon of kerosene on top of the kindling and Stove began balefully to burn a little. I learned by experience that it took two cups of kerosene to get his blood circulating in the morning and that he would only digest bark at night. In the summer and spring I didn't care how slow he was or how little heat he gave out. Bob and I were out doors from dawn to dark and we allowed plenty of time for cooking things and all of the wood was dry and the doors were open and there was plenty of draught. But with the first rainy day I realized that Stove was my enemy and would require the utmost in shrewd, cautious handling.

From the first rain, until late spring, across the kitchen in true backwoods fashion, were strung lines and lines of washing, only slightly less damp than when first hung up days and sometimes weeks ago. Those things directly over Stove flapped wetly against me as I cooked, but I dared not take them down for they were the necessary things like underwear and socks which had to get dry before summer. Try turning the chops, and stirring the tomatoes with someone slapping you across the back of the neck with a wet dish towel—you'll get the idea. I was cold all winter—it seemed that I moved around inside of but without direct contact with my clothes, and my skin became so damply chill that put side by side with a lot of clams I would have found them cozy. Another factor was that being so cold kept me running to the outhouse and each trip made me colder and the next trip frequenter. . . . I wondered if Death Valley Pete wouldn't like a pardner. Our spring and summer had been strenuous to the point of exhaustion and I, at least, having read many books about farms and farmers, had looked forward to winter as a sort of hibernation period. A time to repair machinery, hook rugs, patch quilts, mend harness and perform other leisurely tasks. Obviously something was wrong with my planning, for it took me sixteen hours a day to keep the stove going and three meals cooked. I leaped out of bed at 4 A.M., took two sips of coffee and it was eleven and time for lunch. I washed the lunch dishes and pulled a dead leaf off my kitchen geranium and it was five o'clock and time for dinner. Everyone else in the mountains had dinner at eleven in the morning and supper at five in the evening, but dinner at night was, to me, the last remnant of my old civilized life and I clung to it like a Southern girl to her accent. Even though we had it at five instead of seven-thirty and it was as leisurely as choking down a hot dog at a football game and our conversation consisted chiefly of "Pass the pickles," it was dinner at night.

Another misconception of farm life I had gleaned from books was that winter was a time of neighborliness. In spring and summer we were too tired at the end of the day to do anything but fall into bed, but I imagined that winter evenings would be filled with neighborly gatherings, popping corn, drinking hot coffee, talking politics and crops. How wrong I was. Winter was a time for ordinary chores which took ten times as long to perform because everything was cold and wet and dark and the neighboring farmer's one idea was to get the damn things done so he could go in where it was warm and stay put. The farmer's wife followed the same pattern for winter working that I did, which occupied her for twelve to eighteen hours a day and was roughly as follows:

Monday—Washday! Washing was something that the mountain farm women had contests doing to see who could get it on the line first Monday morning. All except me. I had a contest with myself to see how long I could put off doing it at all. I attacked my washing with the same sense of futility I would have had in attempting to empty the ocean with a teaspoon. Bob had been a Marine in World War I and instead of being shell-shocked he carried home a fixation that a helmetful of water was enough to wash anything, including blankets, and on Monday morning he would say cheerfully at breakfast, "Going to wash today?" and I would answer hopefully, "Yes, and it's going to be a HUGE ENORMOUS washing!" And so Bob would go whistling down through the orchard to the spring and bring back about four tablespoonfuls in the bottom of each bucket and then disappear into the woods where he remained incommunicado until lunch. A few times I left the washing until after lunch but learned that a sufficiency of water does not compensate for having to straddle clothes baskets and wash boards while cooking dinner or having to leave the warm house and hang out wet clothes in the dark. So I carried 99 per cent of my wash water and if I was able to get it hot and could scrub the clothes clean they never dried in winter, so what?

Also the water was so hard it should have been chipped out of the spring and even when mixed 40-60 with soap produced nothing but a greasy scum and after a day spent scrubbing clothes in that liquid mineral I could peel the skin off my hands like gloves.

I entered all of the soap contests in the vain hope that I

would win $5000 and never have to use theirs or any other washing powder again as long as I lived. I failed to understand why farm wives were always talking about the sense of accomplishment they derived from doing a large washing. I would have had a lot more feeling of accomplishment lying in bed while someone else did the washing.

Tuesday—Ironing! Ironing with sad irons has nothing at all to do with preconceived ideas about ironing. It is a process whereby you grab a little portable handle and run over to the stove and plug it into an iron which is always covered with black. Then you run back to the ironing board and get black on your clean pillowcase. You take the iron over to the sink and wipe it off and it is of course too cool, by now, to do any good to the dirty pillowcase so you put it back on the stove and repeat the process until your husband comes in and wants to know where in HELL his lunch is.

Bob was irritatingly casual about my washing and ironing and was continually putting on clean clothes, when he could get them away from me. I got to be just like a dog with a bone over anything I had washed and ironed. It wasn't that I wanted him to act like the advertisements and come dancing into the kitchen in his underwear clutching a clean shirt and yelling "No tattle-tale gray this week, little Soft-hands!" It was just that I wanted him to be conscious of the fact that it took a terrific amount of back-breaking labor to keep us in clean clothes and occasionally to comment on it. "Heaven knows," I would say in exasperation, "you expect and get praise for your work—acting like you delivered every egg with high forceps." I was that way on winter Mondays and Tuesdays—it all seemed so futile.

Wednesday—Baking Day! Each Wednesday plunged me headlong into another, great, losing battle with bread-baking. When I first saw that fanatically happy look light up

Bob's face when he spoke of chickens and realized that this was his great love, I made up my mind that I would become in record time a model farm wife, a veritable one-man-production line, somewhere between a Grant Wood painting, an Old Dutch Cleanser advertisement and Mrs. Lincoln's cookbook. Bread was my first defeat and I lowered my standard a notch. By the end of the first winter, in view of my long record of notable failures, I would probably have had to retrieve this standard with a post-hole digger.

To begin with, the good sport, so the mountain legend has it, made her own yeast by grinding up potatoes, using ONE DRY yeast cake PER MARRIAGE; kept the yeast alive by adding potato water and never allowing the yeast bowl to get cool. I had been on the farm a matter of seconds before I saw that the only way I could keep anything consistently warm would be to stuff it down the front of my dress, so I gave up the homemade yeast idea and resorted to deceit and fresh "store-boughten" yeast.

My first batch of bread was pale yellow and tasted like something we had cleaned out of the cooler. I tried again. This batch had the damp elasticity of the English muffin that tasted like something we had intended to clean out of the cooler but was too heavy.

At Bob's gentle but firm insistence I took a loaf, still quivering from the womb, to a neighbor for diagnosis. Unfortunately, the neighbor, Mrs. Kettle, was just whipping out of the oven fourteen of the biggest, crustiest, lightest loaves of bread I had ever seen. I put my little undernourished lump down on the table and it looked so pitiful among all those great bouncing well-tanned beauties that I had to control a strong desire to jerk it up, nestle it against me protectively and run the four miles home.

Mrs. Kettle had fifteen children and baked fourteen loaves

of bread, twelve pans of rolls, and two coffee cakes every other day. She was a very kind neighbor, a long-suffering wife and mother and a hard worker, but she was earthy and to the point. She picked my stillborn loaf from the table, ripped it open, smelled it, made a terrible face and tossed it out the back door to her pack of mangy, ever hungry mongrels. "Goddamn stuff stinks," she said companionably, wiping her hands on her large dirty front.

She moved the gallon-sized gray granite coffee pot to the front of the stove, went into the pantry for the cups and called out to me, "Ma Hinckley had trouble with her bread too when she lived on your place." I brightened, thinking it might be the climate up there on the mountains, but Mrs. Kettle continued. "Ma Hinckley set her bread at night and the sponge was fine and I couldn't put my finger on her trouble till one day I went up there and then I seed what it was. She'd knead up her bread, build a roaring fire and then go out and lay up with the hired man. When she got back to the kitchen the bread was too hot and the yeast was dead. Your yeast was dead too," she added.

Having quite obviously been given the glove, I hurriedly explained that we had no hired man and the barn was now a chicken house. Mrs. Kettle heaved a sigh for all good things past and poured our coffee. With the coffee she served hot cinnamon rolls, raspberry jam and detailed accounts of the moral lapses of the whole country. It was almost noon when I left for home, clutching a loaf of Mrs. Kettle's bread, two pocketfuls of anecdotes for Bob and a few hazy instructions for myself.

On the long walk home I attempted to strain my baking formula from the welter of folklore but, from that day forward, my wooden bread bowl was to me a sort of phallic symbol and as I kneaded and rolled the unwilling dough I mulled over the little unconventionalities of my neighbors and wondered through which window the hired man used to beckon to Ma Hinckley.

Mrs. Kettle had told me that I didn't work fast enough. That "store-boughten" yeast should never be set the night before and the bread had to be made quickly in one morning. I worked like a frenzied maniac and I baked three loaves of bread twice a week and it made the house smell peasanty and in my letters home I referred very often to my homemade bread, but Bob's reaction—standard—was the true criterion of my success. He said only, "Will it cut?"

Tuesday and Wednesday were also optional bath days. Saturday was a must bath day but because of fires all day for ironing and baking we also took baths on Tuesdays and Wednesdays. This cutting down from daily bathing to a maximum of two complete baths a week wasn't at all hard for me nor for anyone else who has ever taken a bath in a washtub. Washtub baths are from the same painful era which housed abdominal operations without anæsthetics, sulphur and molasses in the spring, and high infant mortality. Both Bob and I are tall—he six feet two inches—and even with conditions right, Stove going, the water warm and scented, towels large and dry (always large and slightly damp in winter) the fact remains that the only adult capable of taking a bath in a washtub in comfort is a pygmy.

A sponge bath in the sink was no sensual orgy either but it was quicker and got off some of the dirt.

Thursday was SCRUB Day! Window washing, table leg washing, woodwork washing, cupboard cleaning in addition to the regular floor scrubbing. I indulged, somewhat unwillingly, in all of these because Bob, whom I accused of having been sired by a vacuum cleaner, was of that delightful old school of husbands who lift up the mattresses to see if the little woman has dusted the springs. I didn't dare write this to Gammy; she would have demanded that I get an immediate divorce. I didn't really object too strenuously to Bob's standards of cleanliness as he set them for himself as well, and you could drop a piece of bread and butter on his premises, except the chicken houses, and I defy you to tell which side had been face down. There was just one little task which brought violent discord into our happy home. Floor scrubbing. By the end of that first winter I vowed that my next house would have dirt floors covered with sand. In the first place Bob had chosen and laid, with great precision and care, white pine floors. Another type of floor which might possibly get as dirty as white pine, or more quickly, would be one of white velvet. Bob was very thoughtful about wiping his feet but he might as well have hiked right through the manure pile and on into the kitchen. I scrubbed the floors daily with everything but my toothbrush, yet they always looked as if we had been butchering in the house for the past four years. Advice from neighbors had been to use lye, but as many of these lye prescribers were missing an eye or portion of cheek—which tiny scratch they laughingly said they got from falling in or over the lye bucket—I filed lye away as a last resort.

I heartily resented having to scrub my floors every day. I thought it a waste of valuable time and energy and accomplished nothing for posterity. I didn't see why beginning with the rainy season we didn't just let the floors go, or cover them with cheap linoleum. But no, mountain farm tradition and Bob's vacuum cleaner heritage had it that I should scrub the floors every day—it was a badge of fine housekeeping, a labor of love and a woman's duty to her husband. The more I was shown of that side of the life of a farmer's good wife, the more I saw in the life of an old-fashioned mistress. "Just don't let anyone tempt me on a linoleum floor." I would growl balefully at Bob.

Friday—Clean lamps and lamp chimneys! I have heard a number of inexperienced romantics say that they prefer candle and lamp light. That they purposely didn't have electricity put into their summer houses. That (archly) candle and lamp light make women look beautiful. Personally I despised lamp and candle light. My idea of heaven would have been a ten million watt globe hung from a cord in the middle of my kitchen. I wouldn't have cared if it made me look like something helped from her coffin. I could see then, and candles could go back to birthday cakes and jack-o'-lanterns and lamps to the attic.

In the first place you need a set of precision instruments and a hair level to trim a lamp wick. Even then it burns straight across for only a moment, then flares up in one corner and blackens the chimney. It's a draw whether you want to use half your light one way or the other—either with the wick turned up and one side of the chimney black or the wick turned below the light line. According to Sears, Roebuck the finest kerosene lamp made only gives off about 40 watts of light so you're a dead cinch to go blind anyway, according to Mazda.

Candle and lamp light are supposed to make your eyelashes look long and sweeping. What eyelashes? Most of the time my eyelids were as hairless as marbles from bending over the lamps to see why in hell those clouds of black smoke.

Saturday—Market Day! In winter Bob left for "Town" while it was still dark, to sell the eggs, buy feed and groceries, get the mail, cigarettes and some new magazines. In spring and summer I joyfully accompanied him, but in the winter driving for miles and miles in a Ford truck in the rain was not a thing of pure joy and anyway, in view of the many ordinary delays such as flat tires, broken springs, plugged gas lines, ad infinitum, I had to stay home to put the lights in the chicken house at the first sign of dusk.

Some Saturday mornings, as soon as the mountains had blotted up the last cheerful sound of Bob and the truck, I, feeling like a cross between a boll weevil and a slut, took a large cup of hot coffee, a hot water bottle, a cigarette and a magazine and *went back to bed.* Then, from six-thirty until nine or so, I luxuriated in breaking the old mountain tradition that a decent woman is in bed only between the hours of 7 P.M. and 4 A.M. unless she is in labor or dead.

Along about three-thirty or four o'clock on Saturday I had to light the gasoline lanterns—the most frightening task on earth and contrary to all of my early teachings that anyone who monkeys around gasoline with matches is just asking for trouble. I never understood why or how a gasoline lantern works and I always lit the match with the conviction that I should have first sent for the priest.

Bob patiently explained the entire confusing process again and again, but to me it was on the same plane with the Hindu rope trick, and it was only when he was not home that I would tolerate the infernal machines in the same room with me. I used to take them out into the rain to pump them up, then crouching behind the woodshed door I reached way out and lit them. Immediately and for several terrible minutes they flared up and acted exactly as if they were going to explode, than as suddenly settled back on their haunches to hiss contentedly and give out candle power after candle power of bright, white light. With two lanterns in each hand I walked through the complete dejection of last summer's garden, ignoring the pitiful clawings and scratchings of the derelicts of cornstalks and tomato vines shivering in the rain, and hung the lanterns in the great chicken house which instantly seemed as gay and friendly as a cocktail lounge. When the frightened squawks of a few hysterical younger hens had died down, I stood and let some of my loneliness drip off in the busy communal atmosphere.

The floor was covered with about four inches of clean, dry straw; and the hens sang and scratched and made little dust baths and pecked each other and jumped on the hoppers and ate mash and sounded as if they were going to—and did—lay eggs. They were as happy and carefree in November, when the whole outside world was beaten into submission by the brooding mountains and the endless rain, as they were on a warm spring day.

Then I gathered the eggs. Gathering eggs would be like one continual Easter morning if the hens would just be obliging and get off the nests. Cooperation, however, is not a chickenly characteristic and so at egg-gathering time every nest was overflowing with hen, feet planted, and a shoot-if-you-must-this-old-gray-head look in her eye. I made all manner of futile attempts to dislodge her—sharp sticks, flapping apron, loud scary noises, lure of mash and grain—but she would merely set her mouth, clutch her eggs under her and dare me. In a way, I can't blame the hen—after all, soft-shelled or not, they're her kids.

The rooster, now, is something else again. He doesn't give a damn if you take every egg in the place and play handball. He doesn't care if the chicken house is knee-deep in weasels and blood. He just flicks a speck from his lapel and continues to stroll around, stepping daintily over the lifeless but still warm body of a former mistress, his lustful eye appraising the leg and breast of another conquest.

Bob used to say that it was my approach to egg gathering which was wrong. I reached timidly under the hens and of course they pecked my wrists and as I jerked my hands away I broke the eggs or cracked them on the edges of the nests. Bob reached masterfully under the hens and they gave without a murmur. I tried to assume this I-am-the-master attitude, but I never for a moment fooled a hen and after three or four pecks I would be a bundle of chittering hysteria with the hens in complete command.

Bob usually got home from "Town" around five and nothing ever again in all of my life will give me ecstatic sensation as did the first sound of his returning truck. Every few seconds I dashed to the windows to note the progress of the lights and then finally in he came, smelling deliciously of tobacco, coldness and outdoors and with his arms laden with mail, newspapers, magazines, cigarettes, candy and groceries. How we reveled in those Saturday nights, smoking, eating, reading aloud and talking; unless, perhaps, as sometimes happened, I had forgot to order kerosene. Then I squeezed the can and poured all of the lamps together and turned way up the wick of the one lamp with the scant cup of kerosene in it. But the effect of the pale, scant-watted light, the sweating walls and Bob's set mouth and hurt eyes was more than a little as if we were trapped in an old mine shaft. Stove loved situations like that and added to the general discomfort by quickly turning black whenever I lifted his lids, then taking advantage of the murky gloom he would put out his oven door and gouge me in the shins. Bob was never one to scold, but he showed his disappointment in me by leaving the table still chewing his last bite and thrusting himself into bed, to dream, no doubt, of the good old days of wife beating.

Sunday! In the country Sunday is the day on which you do exactly as much work as you do on other days but feel guilty all of the time you are doing it because Sunday is a day of rest.

Sunday mornings I cleaned Stove's suit, taking all of the spots off his vest and coat, and it evidently pleased him for he stewed chickens and roasted meat and even exuded a little warmth. Excited by his compatibility, I would mull over recipes for popovers, cup cakes and other hot oven delicacies but would eventually slink back to deep apple pie, as I could use automatic biscuit mix for the crust and our apples were delicious no matter what I did to them.

Also because of Stove's Sunday attitude I washed my hair on that day and guided by the pictures in Saturday's magazines would try the latest hair-dos. Unfortunately my hair is heavy and unmanageable and my attempts at a pompadour usually ended up looking like a tam-o'-shanter suspended over one eye. It made little difference, though, except as a diversion for me, because presently Bob would come in from the chicken house and look hurt and I would put my hair back the old way. I believe that Bob's mother must have been frightened by a candy box cover while she was carrying him, because he wanted me to wear long hair done in a knot, the color blue and leghorn hats, all of the time.

By one o'clock on winter Sundays the house was shining clean, my hair was washed, Bob had on clean clothes and dinner was ready. Usually, just as we sat down to the table, as if by prearranged signal, the sun came out. True it shone with about as much warmth and lust as a Victorian spinster and kept darting behind clouds as if it were looking for its knitting and sticking its head out again with an apologetic smile, but it was sun and not rain. The mountains, either in recognition of the sun or Sunday, would have their great white busts exposed and I expected momentarily to have them clear their throats and start singing *Rock of Ages* in throaty contraltos.

For the few minutes on Sunday when I was not within actual striking distance of Stove or the sink I was wiping up the mud I had tracked in from the woodshed. Bob would be occupied for two full shifts just chopping wood and carrying water. After dinner we would indulge ourselves by grading and packing eggs.

Winter day succeeded winter day and winter week succeeded winter week and the only thing that varied was the weather. No wonder the old timers looked so placid—they didn't have a damn thing to mull over. The days slipped down like junket, leaving no taste on the tongue.

5

Infiltration

WHATEVER my original attitude was, I became reconciled to certain things as unavoidable chuckholes in my road of living on a chicken ranch and grew to accept placidly certain other things, which at first had called for hyperboles of enthusiasm, as just everyday smooth places.

One of the worst chuckholes was getting up at four o'clock in the morning. I got used to it but I felt so strongly about it that many mornings I wondered aloud if I would have married Bob if I had known that this went along with him. He used to laugh at me and swear that he told me but I think it as unlikely as to have courted me with, "And another wonderful thing, dearest, an old prostitute friend of mine is going to live with us." I found that an alarm clock going off at my head at 4 A.M. did nothing toward awakening me—it merely produced shock and when I recovered I was sleepier than ever. I learned that the only solution was to leap from bed at the first jangle, throw on my clothes—and then there is one thing to be said for an outhouse, a brisk walk the first thing in the morning does wake you up.

Bob didn't mind getting up. In fact, he seemed to enjoy it and was odiously cheerful. When the dark winter mornings came around and the rain seemed to be pushing the roof down on us, Bob generously offered to get up and build the fires while I lolled in bed. Of course I accepted and the first morning he tumbled out, managing to untuck the covers on my side and to admit great draughts of chill air between the sheets. "Get some more sleep," he said loudly as he stamped and grunted into his clothes. "I'll soon have a fire." And so I burrowed down into drowsiness and warmth and thought "I have married, without a doubt, the most wonderful man in the world." But I had reckoned without Stove. Suddenly I was ripped from unconsciousness by the crashing of stove lids and my teeth rattling to the rhythm of the ash shaker. This was quickly followed by billows of black smoke and a stream of curses predominated by roars of "Big Black Bastard." When I hurried to the rescue, Bob was amazed and said innocently, "No need for you to get up. Should have slept until I got the fire going." I refrained from stating that it would have been stretching a point to ask a person to stay dead in that racket. From then on I continued to arise and cope with Stove myself.

Definitely a smooth place was the food. I accepted as ordinary fare pheasant, quail, duck, cracked crab, venison, butter clams, oysters, brook trout, salmon, fried chicken and mushrooms. At first Bob and I gorged ourselves and I wrote letters home that sounded like pages ripped from a gourmand's diary, but there was so much of everything and it was so inexpensive and so easy to get that it was inevitable that we should expect to eat like kings. Chinese pheasant was so plentiful that Bob would take his gun, saunter down the road toward a neighbor's grain field and shoot two, which were ample for us, and come sauntering home again. At first under Bob's careful guidance I stuffed and roasted them, but finally I got so I ripped off the breast, throwing the rest away, and sautéed it in butter with fresh field mushrooms. It made a tasty breakfast. The blue grouse were also very plentiful, but the salal berries which they gorged on gave them an odd bitter taste which neither Bob nor I cared for. Quail were everywhere but they were such tiny things that we finally passed them up for the ruffled grouse and the pheasant. There were literally millions of wild pigeons in the valleys. They descended in white clouds when the farmers planted grain and in actual self-defense they shot them even though they were protected by Federal law. Our neighbors gave them to us by the dozens and they were simply delicious, all dark meat and plump and succulent from eating the farmer's wheat, barley, oats and rye. I regret to state that their illegality didn't taint the meat one iota for me. Bob is a fine hunter and a good sport and he, at first, lectured the farmers and their sons on the seriousness of their offense in shooting the pigeons, but the first time he was present at grain planting time and saw what they did to the crops, he told me he thought there should be a bounty on them. He never shot one, however, nor admitted that he enjoyed eating them.

Venison we had twelve months a year, both canned and fresh. To the Indians, who comprised a great part of the population of that country, and to the farmers, who were part Indians, deer meat was meat and game laws were for the city hunters who came in hordes every fall to slaughter all of the bucks. Our local any-season-hunters said they killed only the barren does, which were easily distinguished by their color and which were a nuisance. True or false, the Indian hunters went through the woods without as much disturbance as a falling leaf and the only game warden able to catch them would have been another Indian, and so we had an Indian game warden and the other Indians and the farmers continued to hunt whenever they needed meat and we, in the heart of the deer county, had venison the year round.

Bob usually cooked the game. We underwent this little ordeal because he was of the opinion that only he, and perhaps the chef at the Waldorf, knew how to cook game. With venison he used lots of garlic, pinches of sage, marjoram, bayleaf, pepper, salt, hundreds of pots and pans, Worcestershire, celery salt, onion salt, mushroom salt and everything else he could grab with his large floury hands. When the meat was finally in the oven he hovered around the stove getting in my way and complaining about the quality of the wood (that same wood which he had praised so highly to me and with each armful of which he had guaranteed white heat). When at long last with reverent hands he served me a portion of the venison steak, chops or roast, I found that it tasted just like venison and palled after the second week. Canned with small carrots and onions the venison was delicious. The gamey flavor lost some of its identity in the preserving process and, when the jars were opened months later, the deer meat emerged as a savory stew.

Mushrooms grew profusely around the barns of all the neighboring farms and in our fields. Those around the barns reached a diameter of six inches and the buttons were the size of a baby's fist. Those in the fields were smaller but just as sweet and nutty. One of my early lessons in self-reliance from Daddy had been "How to tell a toadstool from a field mushroom." He bought a large, expensive and profusely illustrated set of nature books including one on mushrooms and toadstools, and for several years we gathered specimens and looked them up and were defeated by the mushroom book's vacillating attitude. "Some are poisonous and some are not," it stated vaguely underneath all of the pictures except those of the field mushroom and the Destroying Angel. Under the picture of the common field mushroom it stated, "This mushroom is often mistaken for the Destroying Angel." Under the picture of the Destroying Angel it said, "This toadstool is often mistaken for the common field mushroom." So

Daddy had us gather large quantities of the Destroying Angel (which is so poisonous that even breathing the spores is dangerous) and large quantities of the common field mushroom in all sizes, from the tiniest button to the full grown kinds, and learn to tell them apart at a glance. We must have learned, because we are all alive and all ardent mushroom gatherers.

The seafood in the Pacific Northwest is superb. The Dungeness hardshelled crabs are the largest, sweetest most delicately flavored crabs obtainable. In Seattle and Portland markets they were usually from 30c to 75c each depending on size. (Today they are sold by the pound and even the medium-sized crabs cost 85c each.) We bought them from the Indians for one dollar a gunnysack full. We'd go on regular crab sprees—eat cracked crab with homemade mayonnaise well-flavored with garlic and Worcestershire, until it ran out of our ears. Have deviled crab, crab Louis and crab claws sautéed in butter and served with Tartar sauce. We never tired of crab and in summer we went often to Docktown Bay, an exquisite little cove below Docktown which was emptied and filled by the tide, and leaned over the sides of a flat-bottomed boat and with long handled nets scooped the scuttling crabs from under seaweed. They didn't compare with the Dungeness crabs, which are gathered from deep icy water, but they were wonderful when boiled on the beach and eaten warm.

This small bay also supplied us with clams—either the large delicately flavored butter clams, which we dipped in flour after removing the neck and stomach, and fried in butter, or the tiny but stronger flavored Little Necks, which we steamed and ate by the bushel. Docktown Bay was a small, warm, horseshoe-shaped cove with a tiny island, about two hundred and fifty feet long by fifty feet wide, in the center of the bay like a dark green dot. When the tide went out it emptied the bay and you could walk to the island and scramble up its steep sides, climb to the top of one of its many trees and feel like Marco Polo. It was a child's island—just the right size and with the slowly returning tide supplying excitement without actual danger. Nearly always, after a picnic, one or two children would be trapped on the island by the tide and then Sharkey, an old Indian with a tremendous head, who lived in a shack on the beach across from the island, would untie his boat, clamber in and row slowly across, shouting in a booming voice, "I'm comin'—I'm comin'."

Sharkey gave me my first geoduck. I had been hearing about geoducks ever since first coming to Seattle. People spoke of them with the mystic reverence usually associated with an eclipse of the sun or the aurora borealis. I had heard that geoducks are giant clams and, like dinosaurs, now extinct. I had heard that geoducks have to be dug by flashlight at night. I had heard that they moved like greased lightning, opening and closing their great shells like a clamshell dredge and getting down to the bowels of the earth in a matter of seconds. I had heard that geoducks have to be dug by a crew of strong men all armed with shovels and all working like demons. I had heard that the only way to catch a geoduck was to take a hatpin and pinion his neck and then excavate under him. I had heard that geoducks were worth driving hundreds of miles and digging all night to get, for they were the most subtly flavored, most succulent of all seafood.

Geoduck, by the way, is pronounced by the Indians gooey-duck.

I had heard all of this about geoducks but like most of the people living in the Pacific Northwest I had not taken the trouble to find out about geoducks for myself, and was therefore shivery with anticipation when Sharkey told me that he had one for me over at his shack, and went over to fetch it. He came back presently and handed me the largest clam I had ever seen. The shell was oblong, about eight inches long and five or six inches wide, squarish and covered with a yellowish skin. The siphon, or what we commonly call the neck, was about two inches in diameter, seven or eight inches long and covered with a heavy wrinkled yellowish skin. The whole thing, including the shell, must have weighed five or six pounds and was definitely unlovely. Sharkey told me to peel the neck and to grind it up for chowder. He said to clean the body part of the stomach and things, cut it into steaks and fry them in butter.

I did as he told me and found that geoduck had been vastly overrated. It was tougher than tire casing and tasted exactly like clams.

I asked Sharkey if he dug it at night by lantern—if he used a hatpin—if geoducks moved fast. He scoffed. "Me dig him low tide this morning. Walk along head down and see big neck in sand. Take shovel and dig like hell and when got big hole I got geoduck. Neck very long and he stick it up to look around and see me comin' and pull it down. He not move, just pull long neck down to where he hidin'. Neck too big to fit in shell, poor thing."

There went all the rumors I had heard—phffffff! There are still people, however, who insist that they have seen geoducks go burrowing into the sand like dredges, ten feet per minute. The Indians all corroborated Sharkey's theory and, as the Indians all got geoducks, it was good enough for me.

Geoducks are protected by game laws and there is a bagging limit on them. One geoduck per person per lifetime would be all right with me. If they were very plentiful, which they are not, and very easy to get, which they are not, geoducks would be very handy for chowder because it would mean cleaning and grinding but one clam and having enough for chowder for an army.

Bob and I, by the way, were of the milk, bacon, green pepper, parsley, potato and onion school of clam chowderers. We never put tomatoes or vegetables in with clams.

Another form of seafood of which we were fond and which we could get for nothing was the oyster. By driving fifty or so miles we could gather both the large soup oysters and the tiny cocktail oysters by the bucketful. The first time we went oystering I was sure there was something wrong and we would all end up in the penitentiary. It didn't seem reasonable to me that oysters, particularly the exquisite little oysters, should be scattered over the countryside free for the picking, and I feared that the shifty-eyed Indian friend of Bob's who was leading the way, would probably next tell us that he knew where there was a great big patch of filet mignon that nobody owned. We drove and drove and drove and took logging roads and cow trails and sometimes seemed to cut right through the brush, but finally we came to a stretch of beach obviously known only to God and that Indian and the oysters were as thick as barnacles. We went there again and again and never saw another soul.

Brook trout could be caught in the irrigation ditches—about ten an hour and from seven to nine inches long. The trout were so thick in the mountain streams that the men in the logging camps caught them with strings, bent pins and hunks of bologna.

The Steelhead salmon came up the irrigation ditches and by trolling near Docktown Bay we caught silvers, king and dog salmon. We also caught sole, cod, red snapper and flounder. Like a true squaw I learned to clean, skin and fillet fish while Bob smoked, looked on and criticized. I really enjoyed it though. I had a knife with a saw edge and a pair of pliers and I could clean, skin and fillet five flounder and two cod while Bob was putting the boat away. It was fun to straddle a log in the warm sunlight, throw the entrails to the gulls and wipe my hands on seaweed. Seaweed reminds me that I suffered one bitter disillusionment in regard to seafood. I had heard for years that clams, freshly dug and steamed in seaweed were the last word in gustatory delight. Accompanied by sweet corn and hot coffee they were too marvellous to bear description. This is a lie. Clams freshly dug and steamed in seaweed are full of sand and unless you are bent on polishing your own fillings there is nothing you can do with them but throw them away. Corn steamed in seaweed is all right if you don't mind boiling juice and seawater oozing up your sleeves. I think that the whole thing should be dispensed with and the food cooked at home. Anyway clams should be soaked over night in fresh water with corn meal in it, so that they can open up and expel the sand.

We had fried chicken for breakfast, lunch and dinner. We had chicken roasted, fricasseed, stewed and in soup and salad. We did not tire of it nor of eggs, but we got damn sick of sowbelly, which is the only meat the local stores ever carried. It was white and fat and could be eaten when seasoned with garlic, salt and sage, but to eat it every day, fried with potatoes, boiled with cabbage or lying sluggishly in a heavy milk gravy as did most of the farmers was not within our capabilities.

Our garden, also, produced lavishly. The soil, a deep brown loam, and the continual rain made holding back rather than forcing the problem. It seemed to me that things sprouted, bloomed, bore, withered and died before I could run into the house and get a pan to pick them in.

With all of the natural resources in the way of food and the ease with which you could grow anything and everything, I never in all of the time I lived on the chicken ranch tasted salad in anyone's house but my own; nor did I see meat cooked any way but fried or boiled, nor did I ever catch anyone but the Indians eating fish. Sowbelly, fried potatoes, fried bread, macaroni, cabbage or string beans boiled with sowbelly were the fare day in and day out. They grew heads of lettuce the size of cabbages and fed it to the chickens or the pigs, they grew celery as crisp and white as crusted snow and they sold every single stalk. They grew beets like balloons and rutabagas as big as squashes, but they fed them to the cows. They grew Swiss chard three feet high, so they cut off all of the green part and fed it to the pigs and boiled the white stems with sowbelly for hours and hours and hours, until it was a greasy strangled mass which they relished with fried potatoes and boiled macaroni.

We could have kidneys, sweetbreads or liver for the asking —"We don't eat guts," the farmers said. We did whenever we could get them. Lamb's kidneys or veal kidneys sautéed in butter, then simmered gently with fresh basil, marjoram, and a wineglass or two of sherry didn't taste like guts to us. Sweetbreads creamed with fresh mushrooms bore little resemblance to guts either. But sowbelly looked and tasted exactly like its name.

Along with the smooth eating was a definite chuckhole. We had breakfast at five in the morning and dinner at five in the evening. Seven and seven would have been bearable, eight and eight enjoyable and nine and nine divine, but five and five it was and I always felt that those meals were like premature babies and lacked the finishing touches.

Another smooth place, which I came to expect as my just deserts, was the scenery. I watched mornings turn pale green, then saffron, then orange, then flame colored while the sky glittered with stars and a sliver of a golden moon hung quietly. I watched a blazing sun vault over a mountain and leave such a path of glory behind that the windows of mountain homes like ours glowed blood red until dark and even the darkness was tinged and wore a cloak of purple instead of the customary deep blue. Every window of our house framed a vista so magnificent that our ruffled curtains were as inappropriate frames as tattered edges on a Van Gogh. In every direction, wherever we went we came to the blue softly curving Sound with its misty horizons, slow passing freighters and fat waddling ferries. The only ugliness we saw was the devastation left by the logging companies. Whole mountains left naked and embarrassed, their every scar visible for miles. Lovely mountain lakes turned into plain ponds beside a dusty road, their crystal water muddy brown with slashings and rubbish.

I loved the flat pale blue winter sky that followed a frosty night. I loved the early frosty mornings when the roofs of the chicken houses and the woodshed glowed phosphorescently and the smoke of Bob's pipe trailed along behind him and the windows of the house beamed at me from under their eaves and Stove's smoke spiral thinly against the black hills. I loved those things but there were the others:

Reading by the wick when I forgot to order kerosene.

Hurling myself headlong through the nearest aperture at the sound of a car but never being in time to tell who it might have been.

Telling time by the place where the sun should have been when I forgot to wind the clock.

Knowing that if I forgot to order matches I would darn well have to learn to rub two sticks together or walk four miles on the loneliest road in the world to a neighbor.

Being asked in the same trustful I-know-you-will-do-it way to split shakes, help fell a tree, dissect a dead chicken, help castrate a pig, run and get the .30-.30 or the shotgun or the .22, flush a covey of quail, retrieve a grouse, wind a fish pole, or make another try at the damned lemon pie which I know that neither Bob's nor any other mother ever made.

Knowing that it was stupid but excusable to run out of food, toilet paper, matches, wood, kerosene, soap or water but that there was *no excuse ever* for running out of shotgun shells or chicken feed.

Being lonely all of the time, I used to harbor the idea, as who has not, that I was one of the few very fortunate people who was absolutely self-sufficient and that if I could just find myself a little haunt far from the clawing hands of civilization with its telephones, electric appliances, artificial amusements and people—people more than anything—I would be contented for the rest of my life. Well, someone called my bluff and I found that after nine months spent mostly in the stimulating company of the mountains, trees, the rain, Stove and the chickens, I would have swooned with anticipation at the prospect of a visit from a Mongolian idiot. And if the clawing hands of civilization could only have run a few telephone and light wires in there they could have had my self-sufficient right arm to chop up for insulators.

Feigning pride and delight in Bob's superior marksmanship while the chill night winds whipped my nightgown around my trembling legs, my blood turned to sherbet and my teeth chattered like castanets. This was always preceded by my having to leap from bed and stand without even benefit of a robe, in the open window while Bob pointed out an owl sitting in a snag about twenty-five miles away. As I am so nearsighted I cannot see anything unless it is perched on my shoulder, I endured long painful periods of "There he is on that first limb." "What limb?" "The one on top of that big snag." "What snag?" "The one east of that tallest fir." "What fir?" And on and on until at last I learned to say with my eyes shut and before I had reached the window, "Oh, yes, I see him plainly," and then I'd run for my bathrobe and my coat and my wool socks.

Owls were worse than hawks for killing chickens and it was fortunate for us that Bob was a crack shot with eyes like telescopes, but it was unfortunate for me that I was not imbued with the thrill of the hunt instead of a hatred for night air and loud noises. I suggested that Bob fire a few shots in the air from the bed through the open window just to let the owl know there was a man in the house, like Gammy pounding Daddy's shoes on the floor, but he gave me a withering look and we dropped that subject like a stone.

While I jounced and eased my way along from day to day, Bob sailed along in front of me never once touching the rough spots. He never seemed to be lonely, he enjoyed the work, he didn't make stupid blunders and then, of course, he wasn't pregnant.

6

Mental Block

WHEN you make a complete change in your mode of living, as I did, you learn that, along with the strange aspects of the new life which seep in and become part of you, will come others to which you never become accustomed. Some of the things I never got used to were:

The hen.

The gasoline lantern.

The outhouse at night where I had a horrible choice of either sitting in the dark and not knowing what was crawling on me or bringing a lantern and attracting moths, mosquitoes, night hawks and bats.

No radio.

No telephone.

Bats hanging upside down in the cellar, flying in the open bedroom windows on summer nights, swooping low over the bed, almost touching my face and making my skin undulate in horror.

Dropping boards and chicken lice.

The inconsistency of a Mother Nature who made winter so wetly, coldly, soggily miserable that I wanted to get back under my stone, and spring so warm, so lush and fragrant that I wanted to roll on my back and whinny.

Rhododendrons being wild. Rhododendrons are expensive shrubs usually grouped in bad color combinations in front of white houses—the ugly purple ones are often banked in front of mustard stucco houses. That is all I knew about the rhododendron until I moved to the mountains. Then I learned that the rhododendron is the native flower of the State of Washington and in the spring and early summer every roadside, every mountainside, every woodsy place on the coast is ablaze with them. They are all pink but range from deep cerise buds to pale pink full-blown blossoms. In the open fields around Docktown Bay, the bushes were about four feet high, rounded in the orthodox bush shape and solid with blooms—the flower heads as big as cabbages—the individual flowerets like single roses. The shiny, dark green foliage, like laurel, was beautiful in its own right. In the thickets and in the woods the long slender rhododendron branches reached as high as twenty feet in an effort to find light and sun and, as they bore their flowers on the uppermost tips of these branches, it was no surprise to walk through the woods, look up and see a lovely cerise bud peering down from the top branches of a small fir or with flushed cheek laid on the cool oily smoothness of a cedar frond. The rhododendrons were so gorgeous, so showy, it was hard for me to believe that they were wild flowers and, as I climbed into trees and scrambled over stumps to get a wonderful armful of buds, I guiltily looked for NO TRESPASSING signs. There is a law that they cannot be picked within fifty feet of the highway, but no one should object to that as they are far prettier and deeper-colored back in the woods.

Rhododendrons grew profusely all over the mountains near our ranch but there were none in the ranch yard, so I asked the man at the feed and seed store in Town how to transplant them. He said to transplant them while they were in bloom and be sure to put them in a shady place where the soil was damp and acid. What he did not tell me was that they have a tap root as big as my arm and about a mile long. Armed with my spading fork and a desire to fill up an ugly corner by the back porch where the soil was so sour it grew only moss, I went out one warm spring evening and found myself three stocky well-formed plants just coming into bloom. I confidently thrust the fork into the soil, making a circle around the base so that there would be a ball of earth on the roots. When everything was well-loosened I slipped the fork under the roots, grabbed the main stalk and nothing happened. The thing wasn't even budgeable. I dug deeper and pried harder and finally located the tap root which went straight down into the bowels of the earth, then took a sharp turn and headed north for a mile or so. With night upon me and the realization that I was going to have to tunnel under the road staring me in the face, I went home and got the hatchet and chopped off the tap root about a foot from the main stem. The next two I chopped before even digging and stuffed them all into a pit I had previously dug by the porch. They not only lived—they throve.

The astonishing fact that there was always on my pantry shelf a water bucket of double-yoked and checked eggs to do with as I would was a source of constant delight and lured me into trying many of the rich, eggy old-fashioned recipes in Mrs. Lincoln's cookbook. In town where I would have had to buy my groceries and balance a food budget, I wouldn't have put up with Mrs. Lincoln and her "beat the whites of sixteen large eggs with a fork on a platter," and her "two wineglasses of old brandy and a cup of slivered, blanched almonds," for two minutes. Mrs. Lincoln was the type who couldn't cook oatmeal mush without adding a flagon of cherry flip and a soupçon of betel nuts. I would have loved to visit Mrs. Lincoln, but she was hell to cook for unless you lived on a chicken ranch, and then you and Mrs. Lincoln could see eye to eye about a lot of things. Particularly eggs. I had already made sunshine cake, angel food and pound cake and was wondering what would be good on a rainy wet winter day when I chanced on cream puffs. "Now there *is* something," I said, for cream puffs were an old favorite of mine and they used lots of eggs. The recipe called for "eight eggs to be broken one by one and beaten into the mixture with the bare right hand."

"Now, Mrs. Lincoln, let's not be *frugal!*" I said and used sixteen eggs. This made gallons of dough and almost broke my arm but if Mrs. Lincoln could do it, at her age, so could I. "Put pieces of dough the size of walnuts in the pan, leaving *plenty of room,* as they will puff to the size of large apples." I did but when I took them out of the oven they were still the size of walnuts but as hard as diamonds. Down but not out, I got out my deep fat kettle. When the fat was smoking hot I dropped in a piece of the dough. Poufffff—the little thing swelled to the size of a cantaloupe. I was ecstatic. For hours I dropped little walnuts into the fat and pulled out great, golden puffs. Then sweating but happy I whipped a large bowl of canned milk. "We'll each fix our own," I said proudly to Bob as I put them on the dinner table and hurried back for the canned milk. I cut mine open to put in the filling but it was already filled—filled with cold grease. They all were, and not only that, but whipped canned milk, in case you didn't know, tastes exactly as burning rubber smells.

I never became acclimated to discussions of wire worms, intestines, chicken lice, ad nauseam at the breakfast table. I used to think that Bob's finer feelings would make a good emery bag as I watched him closely examine a colored diagram of a chicken's wormy insides, then with relish take a spoonful of soft boiled egg, back to the diagram, then to the egg. I would nervously sip my coffee and try to concentrate on last week's paper.

Even when I was clear and away from the kitchen and feeling in first-class shape, not pregnant, in order to take care of the various ills of or to perform a post-mortem on a chicken, I would have to say over and over, "This is just like being a doctor! This is just like being a doctor!" And sometimes I was sick and sometimes I was not. But it took me several days to get back to eggs again. Bob said that it was in the mind but I reminded him that cleaning fish made him sick and I had to take the dead rats out of the attic. There is no explaining those things. A large husky Swedish farmer down the valley was known to be the best butcher in the country. One day during the fall, Bob went down to see him about butchering our pigs. Mr. Larsen was in the barnyard and while Bob talked to him he knocked on the head, slit the throats and eviscerated two calves without twitching a muscle, but when his wife cut her hand on the separator, a few minutes later, Bob had to stop the flow of blood and bandage it while Mr. Larsen turned green. Mopping his moist forehead with a handkerchief held by a hand and arm, red to the elbow with the calves' blood, Mr. Larsen said, "Blood always did make me sick."

I found it impossible to remember that almost everyone was part Indian. I commented on this to a tall blond woman named Selma Johnson whom we picked up on the road one day and drove to the Docktown store. She laughed heartily and said, "Don't let it bother you. Now, I'm one third bow and arrow, myself. Dad's a Swede and Mom's an Indian and I look like a Swede and my older sis looks like Pocahontas. The only thing I inherited from Mom was good teeth. All us Indians got good teeth," and she laughed again, exposing her milk-white perfect teeth.

I learned that first year that I must not be embarrassed or incensed at the most personal questions. In a country where breeding, fertility and birth were of prime importance in livestock and were discussed casually all the time, it stood to reason that breeding, fertility and birth in humans, though not so important, would be discussed as casually. I turned crimson the first time a farmer, almost a total stranger to me, leaned across the supper table and said to his wife, "Vera, tell Betty about the time you miscarried when we had the preacher to supper." I grew used to it though, as I grew used to all the food being boiled and the homes of the illegitimate sons of the illegitimate sons of the illegitimate sons dotted over the countryside like naturalized bulbs. The legitimate and otherwise gathered together for holidays and anniversaries and no one seemed ashamed of his relationship. There's no use crying over spilt milk, we of the mountains said.

Pregnancy was referred to as being "that way." My being "that way" went the rounds of the mountains and valleys along with the news about the contagious abortion in the Helwig herd of Jerseys and the impotency of the Green bull. One day when Bob and I were driving to Town a man hailed us. We stopped and he climbed on the running board and leaned into the car confidentially. "Say," he said, "heard you was that way." "Yes," I said, "I am." The man leaned in farther so that his face was uncomfortably close to mine. "Just say the word and I'll fix you up. Drop up some evening with six dollars and I'll fix you good as new. Not a thing to it," he said winking at Bob. "Took care of Mrs. Smith when she was six months along and got rid of three for my own wife at three months. Just a plain old-fashioned buttonhook. Nothing to it."

"Oh, him!" said the girl in the doctor's office in town. "His wife's in the hospital right now recovering from her last abortion. We get his work in here all the time," and she laughed heartily. I didn't think it was funny. "Why don't they stop him? Why don't they arrest him?"

The girl sighed and looked out the window. "If it wasn't him it would be someone else. If they can't find someone else to do it they abort themselves. The hospital's full of 'em all the time. Buttonhooks, baling wire, hatpins. God, they're dumb." Not dumb—pitifully ignorant.

I put up my Christmas tree during the last week of November, just to get the feel and smell of November out of the house. Bob warned me that it would dry out and the needles would fall off before Christmas but I laughed. Not only did I think the drying out improbable but it seemed more likely that it would flourish and give birth to little Christmas trees in the moist atmosphere of the house.

I never tired of admiring and loving our little Christmas trees. When we cleared the back fields, Bob let me keep about ten of the prettiest trees for future Christmas trees. The loveliest of all we sent home to the family but the one I chose for our first Christmas was a dear, fat little lady with her full green skirts hiding her feet and all of her branches tipped with cones. During the summer and fall I used to go out and stroke and smell my little tree and it made me feel guilty to look around at her little brothers and sisters crowding the fence and peering wistfully into the yard. Clearing land there in the mountains was like holding back a mob at a fire. As long as the fences held and we were ever watchful, we were safe enough but one break and the trees surged in. We were constantly pushing them back from the garden, the road, the driveway, the chicken yard; and the mountains were carelessly letting them slide down on us. I expected to look up some day and see a mountain bare shouldered and grabbing frantically for her trees.

The family implored us to spend Christmas with them but we couldn't leave the chickens and so they sent lavish boxes and we retaliated with Sears, Roebuck "multicoloreds" and "floral backgrounds." It rained on Christmas and it differed from other winter days only in that Stove balked and refused to have anything to do with the twenty-two pound turkey Bob had bought, and so we had dinner at ten-thirty at night instead of five. Christmas is best with a large family, I found.

After Christmas it rained and rained and rained and dusk settled like a shroud at a little after three o'clock. From the forlorn grayness of the burn would come the sharp crack of a falling snag. Even when there wasn't a breath of wind the poor old things would release their grip on the earth and fall with a splintering crash. The burn, known always as the Big Burn, extended from our road to an arm of the sea about fifty miles away. It was about five miles wide but gradually grew narrower as the mountains did their best to draw the trees together over this unsightly rip which showed their bare skin through the green.

Years ago fire had swept up this great ravine and was evidently finally checked by the stream whose dry bed was our road. Like a pestilence-struck village, the burn was covered with the gaunt dying bodies of the sick, the fallen rotting bodies of the dead and over everything crawled the marauding blackberry vines, nettles and fireweed. A few low-class squatters, like alder, salal, wild raspberry and blackcaps had made some half-hearted attempts at reclamation but only in the earliest spring were these even noticeable. For some strange reason it was fine hunting ground for birds, rabbits and deer. I hated the burn. In summer it was parched and dry and ugly, and in winter it was gray and soggy and ugly. Mists haunted it day and night and winds came roaring up its entire length and crashed headlong into the house or came crawling from under the vines and logs on their bellies up through the orchard to snivel and whine at the doors and windows. The skyline of the burn was so bleak and hopeless it made me want to run home, light all the lamps and huddle by Stove. In summer the orchard and the alders and maples across the road hid it from the house, but on gray winter days its snaggle-toothed horizon could be seen plainly. Some winter days great winds came bounding down out of the north; blew rain at us in spitty gusts; sent the mountains' misty veils flying, exposing their pale haughty faces; crashed around on the burn, snapping giant snags and tossing terrible handfuls of limbs just anywhere; grabbed our house by the scruff of its neck and shook it until the windows rattled and the shakes flew off; sniffed around the eaves of the chicken house hoping for a loose board; then dashed back to annoy the mountains again, prostrating the small trees in its path. It was boisterous and noisy and terrifying. The only redeeming feature of this terrible wind was that it reached down the chimney, yanked the smoke out and made Stove roar and crackle in spite of himself.

On stormy days I lit the lamps early and stayed close to the house and Stove.

Bob seemed oblivious of the weather. Apparently lulled by the screaming wind, the falling trees, the lashing rain, he whistled gaily as he pumped up his lanterns and began his evening chores. Bundled in oilskins, lanterns swinging like beacons, feed buckets clanking cheerfully, he walked briskly through the rain. He never even noticed the terrible nearness of the mountains.

PART THREE

Spring

Hear ye not the hum of mighty workings!

—KEATS

7

The Whistle Blows

UNTIL I moved to the ranch, the coming of spring had been a gradual and painless thing, like developing a bust. In Butte the snow melted and made torrents in the gutters, the streets didn't freeze at night, we found our first bluebell and it was spring and we could take off our "Chimaloons." In Seattle the seasons ran together like the stained-glass-window paintings we did at school where we wet the drawing paper first, all over, then dropped on blobs of different colors which ran into each other so that it was impossible to tell where one began and the other left off. Seattle spring was a delicate flowering of the pale gray winter—a pastel prelude to the pale yellow summer which flowed gently into the lavender autumn and on into the pale gray winter. It was all very subtle and, as we wore the same clothes the year around and often had beach fires in January but found it too cold for them in June, we were never season conscious.

Things were certainly different up on the ranch. Spring stopped there with a screech of brakes. Somewhere someone blew a whistle and all hell broke loose. We awoke one morning to a new Sears, Roebuck catalogue; baby chickens, thousands of them; a new little red-haired baby girl; little yellow goslings; two baby pigs; a puppy; two kittens; a little heifer calf; fruit trees snapping into bloom all over the place; a newly plowed plot for the biggest garden in the world; streams and lakes brimming; trilliums, wild violets both purple and yellow, camas and starflowers carpeting the woods; fences to mend; seeds to plant; seed catalogues to dream through; Government bulletins to choke down and digest; and no rest ever any more.

The spring sun, a bold-faced, full-blooded little wench, obviously no kin to the sallow creature who simpered in and out occasionally during the winter, bestowed her warm caresses impartially on the handsome virile timber, the tender plowed land and the ugly impotent burn. Every place she touched throbbed hopefully and there was a rapid spreading epidemic of pale green mustaches and beards. The mountains' noses began to run and though they tied veils over their heads they seemed less formidable.

Bob announced that the Big Burn was going to have a bumper crop of wild strawberries, which was like telling me that Dracula was really somebody's mother.

I was so ebullient from the sun and warmth that even the fact that I had to dogtrot through the long days, in order barely to scratch the surface of my thousands of new duties, failed to dampen my ardor.

Bob's spirits, never depressed by the dreary winter rains as mine had been, soared too, and in spite of the fact that for months I had been lifting him out of the abyss of my bad management with the bright hope that I would become marvellously efficient when working under pressure, and unfortunately we both found that like the Government when given more bureaus to handle, I merely became much more inefficient on a much larger scale.

I had read of beauty-starved farm wives standing for an hour on their back stoops absorbing the glory of a sun-drenched branch of forsythia; walking in the orchards and burying their noses in the fragrant boughs; standing motionless in the warm spring sun and thanking God for the miracle of fertility. What I wanted to know was, where they got the time for such ethereal pursuits. I saw the forsythia, I saw the apple blossoms, I saw the sun glancing over the emerald-tipped firs and pointing up the chartreuse maples and alders on its way. I saw those things but I had about as much chance to linger and appreciate as I would have had riding a motorcycle through an art gallery.

It all began with the baby chickens—they came first, while I was still very pregnant, and getting down on my hands and knees to peer under the brooder at the thermometer was a major undertaking. Bob and I scrubbed the brooder house, walls, floors, even the front porch with Lysol and boiling water. The brooder house had two rooms—the brooder room and the cool room. In the brooder room we had two coal-oil brooders which we lit and checked temperatures on, a week before the chicks arrived. The brooder room floor was covered with canvas and peat moss and had drinking fountains and little mash-hoppers scattered here and there. The cool room also had peat moss on the floor and buttermilk and water fountains and mash hoppers here and there. At last the chicks arrived and Bob drove down to Docktown and returned with ten cartons with air holes along the sides, in each of which yeeped one hundred chicks. We stacked the cartons in the cool room and then one by one we carried them carefully into the brooder room, took off the lids and gently lifted out the little chicks and tucked them under the brooder, where they immediately set to work to suffocate each other.

From that day forward my life was one living hell. Up at

four—start the kitchen fire—put the coffee on—go out to the baby chicks—come back and slice off some ham and sling it into the frying pan—out to the baby chicks with warm water—put toast into the oven—out to the baby chicks with mash—set the breakfast table—out to the baby chicks with chick food—open a can of fruit—out to the baby chicks and on and on through the day. I felt as if I were living in a nightmare, fleeing down the track in front of an onrushing locomotive. I raced through each day leaving behind me a trail of things undone. Of course, I chose that most inconvenient time to have the baby and her arrival quite typified the tempo of our life. I rode the fifty-odd miles to town sitting on her head, and the moment I reached the hospital she popped out, red-haired and weighing eight and a half pounds. When I came home from the hospital after two weeks of blissful rest, everything on the ranch had been busy producing and I was greeted by the squealing of baby pigs, the squeaking of baby goslings, the baaing of a heifer calf, the mewing of tiny kittens, the yelping of a puppy and the stronger louder yeeping of the chicks. All of the small eat-often screamers were assigned to my care and I found that feeding of them all and Bob and me was a perpetual task. I relegated my ironing to something I would try and finish before small Anne entered college—my washing I tried to ignore, although it assumed the proportions of a snowball rolled from the top of Mt. Olympus—and I closed my eyes to Spring who was imploring me from every side to do something, anything, about my garden.

Bob's life was as harried, and our marriage became a halloo from the brooder house porch to the manure pile; a call for help when pulling a stump or unrolling some wire; a few grunts at mealtime as we choked down our food and turned the leaves of seed catalogues and Government bulletins. One night after dinner as I sat at the kitchen table industriously making my baby chick "feed and death" entries for the day, Bob unexpectedly kissed the back of my neck. I was as confused as though an old boss had chosen that means of rewarding me for a nice typing job. "Another year or two and we probably won't even use first names," I told Bob.

8

People

THE most important people in a community are usually the richest or the worthiest or the most useful, unless the community, like ours, happens to be scattered thinly over the most rugged mountains and the largest stand of Douglas fir on the North American Continent. Then the most important people are the closest. Your neighbors. Our neighbors were the Hickses and the Kettles.

My first brush with the Kettles came about two weeks after we moved to our ranch and before we had bought our dozen Rhode Island Red hens, when I in my innocence thought I would walk to a neighbor's and arrange to buy milk and eggs. Bob had gone to Docktown after lumber or I probably never would have made that fruitless voyage.

I remember with what care I donned a clean starched housedress and pressed my Burberry coat. How carefully I brushed my hair and fixed my face and composed little speeches of introduction. "So, you're Mrs. Kettle! Bob has told me so much about you!" or "I'm your new neighbor up on the mountain and I thought it about time to come down out of the clouds and make myself known!" (Ha, ha.)

My first disappointment was a little matter of distance. It was possible to keep my spirit of good will and neighborliness whipped to a white heat for about a mile, then it began to cool slightly and by the fourth mile the whole thing had become a damned bore and I wondered why I ever had the idea in the first place. It had rained hard the night before and the road, normally pocked with holes and pits, was dotted with little lakes and pools, which reached clear across the road and oozed into the salal along the edges. In order to traverse these it was necessary to make detours into the soaking wet brush so that by the end of the first mile my neatly pressed coat slapped wetly against my legs and my hair and shoulders were full of twigs and stickers.

The day was clear and blowy with clouds like blobs of thick white lather sailing along on the wind, which was so strong and so playful that incredibly tall, spindly, snags leaned threateningly toward me, particularly when I was trying to edge around an especially large puddle and couldn't have got out of the way if the snag had shouted *"Timbah!"* before it fell. This fear of falling snags wasn't just idle terror on my part either, because every once in a while there would be a big blundering crash to the right or left of me. As the snag was usually just about to hit the ground by the time I had it located, I finally gave up and decided that if God willed it, God willed it, and there was nothing I could do about it.

On either side of the road were dense thickets of second growth, clear green and bursting with health and vigor. Back of these thickets rose the giant virgin forests, black and remote against the sky. Occasionally a small brown rabbit flipped into the bush just ahead of me and little birds made shy rustling noises everywhere. The mountains looked down scornfully at my skip, hop and jump descent and when I saw their unfriendly faces reflected in the puddles I felt the resisting power of that wild country so strongly that I was almost afraid to look back for fear the road would have closed up behind me and there would be nothing but trees, sky and mountain and no evidence that I had ever been there.

Lost in these gloomy thoughts I trudged on until I turned a bend and suddenly came on the Kettle farm. First there was a hillside orchard, alive with chickens as wild as hawks, large dirty white nuzzling pigs and an assortment of calves, cows, horses and steers. Wild roses laced the fences and dandelions glowed along the roadside and over and above the livestock arose the airy fragrance of apple blossoms.

Below the orchard were a large square house which had apparently once been apple green; a barn barely able to peep over the manure heaped against its walls; and a varied assortment of outbuildings, evidently tossed together out of anything at hand. The pig house roof sported an arterial highway sign and the milkhouse had a roof of linoleum and a wooden Two Pants Suit sign. All of the buildings had a stickery appearance, as any boards too long had been left instead of sawed off. The farm was fenced with old wagons, parts of cars, broken farm machinery, bits and scraps of rope and wire, pieces of outbuildings, a parked automobile, old bed springs. The barnyard teemed with jalopies in various stages of disintegration.

I turned into a driveway that led along the side of the house but there arose such a terrific barking and snarling and yapping from a pack of mongrels by the back porch, that I was about to leap over the fence into the orchard when the back door flew open and someone yelled to the dogs to "stop that goddamn noise!" Mrs. Kettle, a mountainously fat woman in a very dirty housedress, waddled to the corner of the porch and called cordially, "Come in, come in, glad to see you!" but as I drew timidly abreast of the porch my nostrils were dealt such a stinging blow by the outhouse lurking doorless and unlovely directly across from it that I almost staggered. Apparently used to the outhouse, Mrs. Kettle kicked me a little path through the dog bones and chicken manure on the back porch and said, "We was wonderin' how long afore you'd git lonesome and come down to see us," then ushered me into the kitchen, which was enormous, cluttered and smelled deliciously of fresh bread and hot coffee. "I'll have a pan of rolls baked by the time the coffee's poured, so set down and make yourself comfortable." She indicated a large black leather rocker by the stove and so I sat down gratefully and immediately a long thin cat leaped into my lap, settled himself carefully and began purring like a buzz saw. As he purred I stroked him until I noticed a dark knot of fleas between his eyes from which single fleas were disentangling themselves and crawling down on to his nose and into the corners of his eyes and then unhurriedly going back into the knot again. I gently lifted him off my lap and put him down by the stove but he jumped back again and I pushed him off and he jumped back and so finally I gave up and let him stay but stopped stroking him and tried to keep track of the fleas to be sure they went back after each sortie.

The Kettles' kitchen was easily forty feet long and thirty feet wide. Along one wall were a sink and drainboards, drawers and cupboards. Along another wall was a giant range and a huge woodbox. Back of the range and woodbox were pegs to hang wet coats to dry but from which hung parts of harness, sweaters, tools, parts of cars, a freshly painted fender, hats, a hot water bottle and some dirty rags. On the floor behind the stove were shoes, boots, more car parts, tools, dogs, bicycles and a stack of newspapers. In the center of the kitchen was a table about nine feet square, covered with a blue and white oilcloth tablecloth, a Rochester lamp, a basket of sewing, the Sears, Roebuck and Montgomery Ward catalogues, a large thick white sugar bowl and cream pitcher, a butter dish with a cover on it, a jam dish with a cover on it, a spoonholder, a fruit jar filled with pencil stubs, an ink bottle and a dip pen. Spaced along other walls were bureaus, bookcases, kitchen queen, worktables and a black leather sofa. Opening from the kitchen were doors to a hall, the parlor, the pantry (an enormous room lined with shelves), and the back porch. The floor was fir and evidently freshly scrubbed, which seemed the height of useless endeavor to me in view of the chicken manure and refuse on the back porch and the muddy dooryard.

While I was getting my bearings and keeping track of the fleas, Mrs. Kettle waddled between the pantry and the table setting out thick white cups and saucers and plates. Mrs. Kettle had pretty light brown hair, only faintly streaked with gray and skinned back into a tight knot, clear blue eyes, a creamy skin which flushed exquisitely with the heat, a straight delicate nose, fine even white teeth, and a small rounded chin. From this dainty pretty head cascaded a series of busts and stomachs which made her look like a cooky jar shaped like a woman. Her whole front was dirty and spotted and she wiped her hands continually on one or the other of her stomachs. She had also a disconcerting habit of reaching up under her dress and adjusting something in the vicinity of her navel and of reaching down the front of her dress and adjusting her large breasts. These adjustments were not, I learned later, confined to either the privacy of the house or a female gathering—they were made anywhere—any time. "I itch—so I scratch—so what!" was Mrs. Kettle's motto.

But never in my life have I tasted anything to compare with the cinnamon rolls which she took out of the oven and served freshly frosted with powdered sugar. They were so tender and delicate I had to bring myself up with a jerk to keep from eating a dozen. The coffee was so strong it snarled as it lurched out of the pot and I girded up my loins for the first swallow and was amazed to find that when mixed with plenty of thick cream it was palatable. True it bore only the faintest resemblance to coffee as I made it but still it had a flavor that was good when I got my throat muscles loosened up again.

As we ate our rolls and drank our coffee Mrs. Kettle told me that she and Paw had fifteen children, the youngest of whom was then ten. Seven of these children lived at home. The other eight were married and scattered in and around the mountains. Mrs. Kettle began most of her sentences with Jeeeeesus Key-rist and had a stock disposal for everything of which she did not approve, or any nicety of life which she did not possess. "Ah she's so high and mighty with her 'lectricity," Mrs. Kettle sneered. "She don't bother me none—I just told her to take her old vacuum cleaner and stuff it." Only Mrs. Kettle described in exact detail how this feat was to be accomplished. As Mrs. Kettle talked, telling me of her family and children, she referred frequently to someone called "Tits." Tits' baby, Tits' husband, Tits' farm, Tits' fancywork. They were important to Mrs. Kettle and I was glad therefore when a car drove up and Tits herself appeared. She was a full-breasted young woman and, even though Mrs. Kettle had already explained that the name Tits was short for sister, I found it impossible to hear the name without flinching. Tits was a Kettle daughter and she had a six-month-old son whose name I never learned as she referred to him always as "You little bugger." Tits fed this baby pickles, beer, sowbelly and cabbage and the baby ungratefully retaliated with "fits." "He had six fits yesterday," Tits told her mother as she fed the baby hot cinnamon roll dipped in coffee.

Then there were Elwin Kettle, a lank-haired mechanical genius, who never seemed to go to school, although he was only fifteen, but spent all of his time taking apart and putting together terrible old cars; and Paw Kettle whom Bob aptly described as "a lazy, lisping, sonofabith." The other Kettles were shiftless, ignorant and non-progressive but not important.

On that first visit Mrs. Kettle told me that she had been born in Estonia and had lived there on a farm until she was fourteen; then she had accompanied her mother and father and sixteen brothers and sisters to the United States and, somewhere en route to the Pacific Coast, had been unfortunate enough to encounter and marry Paw. Immediately thereafter she began having the fifteen children who were all born from ten to fourteen months apart and all delivered by Paw. Mrs. Kettle was plunging into a detailed recital of the conception and birth of each, when I hurriedly interrupted and asked about the milk and eggs. She was shocked. Sell milk? They had never even considered it. They separated all of their milk and sold the cream to the cheese factory. Nope, selling milk was out of the question. "What about eggs?" I asked. "Well," said Mrs. Kettle, "Paw just hasn't gotten around to fixing any nests in the hen house and so the chickens lay around in the orchard and when we find the eggs some are good and some ain't." I hurriedly said that that was all right, I could get the eggs in town, took my leave and went home, and there learned that Bob the efficient, Bob the intelligent, had already arranged with the Hickses for milk and eggs.

Evidently my call was the opening wedge, for the next morning, just after I had finished the breakfast dishes and Bob and I were at work on the pig house, we suffered our first encounter with Mr. Kettle. He came careening into the yard precariously balanced on the top of a flight of steps which formed the seat of his wagon and driving a team composed of a swaybacked stallion about eighteen hands high and a slight black mare little larger than a Shetland pony. Mr.

Kettle drew them to a flourishing halt just as I pictured them charging through the side of the house, and wished us a cheery good morning. Then leaping from the leaning tower of steps to the ground with the air of a Roman charioteer who had just won a race, he stopped and examined his steeds' flanks, did little things to the harness, a masterpiece of ingenuity consisting of baling wire, bits of rope, heavy twine and odd lengths of strap, then straightened up and lit a small piece of cigar. Bob stood transfixed staring at the wagon and team. The small horse staggered under a pair of great brass hames while the stallion wore none; the front wheels of the wagon were easily four feet in diameter and iron, those in back delicate rubber-tired sulky wheels; the wagon itself was the body of a hayrack without the sides and garnished with a flight of steps sloping toward the rear and leading heavenward. I was more fascinated by Mr. Kettle. He had a thick thatch of stiff gray hair quite obviously cut at home with a bowl, perched on top of which he wore a black derby hat. His eyebrows grew together over his large red nose and spurted out threateningly over his deepset bright blue eyes. He had a tremendous flowing mustache generously dotted with crumbs, a neckline featuring several layers of dirty underwear and sweaters, and bib overalls tucked into the black rubber hip boots. Drawing deeply on the cigar butt Mr. Kettle said, "Nithe little plathe you got here. Putty far up in the woodth though. Latht feller to live here went crazy and they put him away." He scrutinized Bob from under his eyebrows. Bob laughed and said, "Well, how do I look?"

Mr. Kettle said, "All right tho far." He turned to me, "The old lady tellth me you wath down yethtiddy. Gueth I mutht have went to town jutht afore you come. Too bad. Too bad." He continued to smoke and we all looked at each other expectantly. Mr. Kettle broke the silence. "Thingth ith putty tough thith year. [We learned the hard way that this was his stock approach to borrowing.] Yeth thir. Tough! The boys WON'T HELP MAW AND ME [his voice seemed to break bounds and rose and fell like the crescendos of a siren] and we can't do it all alone and I GOT TWO THICK COWTH AND WE wondered if you folkth would give uth a hand becauth the boyth are working in the campth in the woodth logging and I CAN'T PLOW ALONE AND THE OLD lady wondered if when you come down YOU WOULD BRING a little kerothene and a little pullet masth, ten cupth of FLOUR AND A FEW RAITHINS if you got 'em." Innocently we agreed to everything and Paw leaped to the flight of steps, clucked to the horses and catapulted out of the yard. From that day forward the flour, chicken feed, eggs, bacon, coffee, butter, cheese, sugar, salt, hay, and kerosene which the Kettles borrowed from us, placed end to end, would have reached to Kansas City—the flour, chicken feed, eggs, bacon, coffee, butter, cheese, sugar, salt, hay, and kerosene which they had already borrowed from the rest of the farmers in the mountains would have reached from Kansas City to New York and back to the coast. There was nothing anyone could do about this borrowing, though. With the nearest store seventeen miles away, you could not refuse to lend someone coffee, flour, eggs, bacon, butter, cheese, sugar, salt, hay or kerosene, because you yourself knew what it was like to run out of any or all of them. Paw Kettle banked on this knowledge and the rest of us charged it off to overhead.

The business of lending our services was something else again, and after that first initial mistake, we seldom if ever granted any of Paw's millions of request for help—help with the plowing, the sowing, the haying, the milking, the barn cleaning, the chicken house building, the gardening, the cess pool, the outhouse moving. He asked and was refused, but he kept right on asking, for that was Paw's business—begging. He didn't care what humiliations, what insults it entailed—it was better than working.

Actually the Kettle farm was the finest, or rather could have been the finest, in that country. They had two hundred acres of rich black soil, of which about twenty, including the acre or so rooted by pigs and scratched by chickens, were under cultivation. Their orchard, which was never pruned or sprayed, bore old-fashioned crunchy dark red apples, greengage plums, Italian prunes, russet and Bartlett pears, walnuts, filberts, chestnuts, pie cherries, Royal Anne's and Bings. Their loganberry, currant, raspberry and blackberry bushes bore with only the spasmodic cultivation given them by rooting pigs and scratching chickens; their thirty-five Holstein cows were never milked on time, rarely fed and beset by flies and vermin but they gave milk, apparently from force of habit; their Chester White sows were similarly abused but they bore huge litters which Paw sold for $5.00 each piglet as soon as they were weaned. Occasionally the Kettle animals just up and died. Such deaths were immediately attributed to a vengeful providence and never for a second did any Kettle entertain the idea that dirt or malnutrition had anything to do with it.

Of course, we didn't know all of those things that next morning when with charity in our hearts we set out to help with the plowing, which we honestly thought Paw Kettle intended to struggle with by himself unless Bob helped him. As we cautiously drove the car through the conglomeration of old cars, parts of old cars, Kettle boys under old cars and discarded furniture which studded the driveway, Paw hallooed down by the barn, so Bob let me off to walk to the house and he drove in the direction of Paw. When I got to the house I found Mrs. Kettle in the throes of cleaning the bathroom and jubilant over an apronful of tools, the top of a still and an unopened package from Sears, Roebuck which had been missing for a year or so. The bathroom was definitely an afterthought tacked on to one wall of, and accessible only through, the parlor. It was just a bathroom, containing a solitary tub and evidently used only through the warm weather. Knowing that they had a good stream, a ram and a water tower, I asked Mrs. Kettle why they didn't install an inside toilet. She was incensed. "And have every sonofabitch that has to go, traipsin' through my parlor? When we start spendin' money like drunken sailors it won't be for no lah-de-dah toilet." I slunk into the parlor and after pulling up the green-fringed blinds I did a little self-conscious dusting under the cold surveillance of rows of "Stony Eyes," Gammy's name for chromo portraits, which lined the walls and were apparently the forebears of Maw and Paw photographed post-mortem. The parlor was clean and neat. The dark red brick fireplace morbidly sported a fern where the fire should have been and from the edge of the mantel were suspended, by tacks and strings, folding red paper Christmas bells, cardboard Easter eggs and greeting cards from birthdays, Valentine's day, Christmas and Easter. At one end of the mantel stood a very bold-faced Kewpie doll clad only in an orange ostrich feather skirt and with no back; at the other end was a much-gilded figurine of the Madonna. The furniture was all slippery black leather; the floor slippery mustard and rust linoleum, and the golden oak library table in the center of the room wore a dung-colored tapestry cover on which were laid at angles a pocketbook of Shakespeare, a mother of pearl encrusted photograph album, a stereoscope and a box of photographs which said on the lid in gold, VIEWS OF YELLOWSTONE NATIONAL PARK. From the lamp hook in the center of the ceiling hung three long curls of flypaper limp with age and heavy with petrified flies. The whole atmosphere was funereal and remote and there wasn't a marred place or a scratch on anything. I was amazed considering the fifteen children and the appearance of the rest of the house. But, when I watched Maw, come out of the bathroom, firmly shut the door, go over and pull down the fringed shades clear to the bottom, test the bolt on the door that led to the front hallway and finally shut and lock the door after us as we went into the kitchen, I knew. The parlor was never used. It was the clean white handkerchief in the breastpocket of the house.

As soon as we finished the bathroom and parlor it was time to get dinner. For dinner we had boiled macaroni—not macaroni and cheese, just plain boiled macaroni without even salt —boiled potatoes, baked beans and pickles, washed down with large white cups of the inky black coffee which had been sulking on the back of the stove since breakfast.

The men gulped their food and hurried back to the plowing. Bob seemed a little grim. Maw and I lingered over the coffee, the lunch dishes and her complaints that her own sisters had been to see her just a few hours before Georgie, Bertha, Elwin, Joe, John or Charles were born and didn't even know that she was "that way." This did not surprise me a great deal as she looked as though she might be going to give birth to an elephant any moment.

About three o'clock Bob appeared and we left rather suddenly. Bob told me through clenched teeth that he had had to stop every five or ten minutes to mend the harness or to scoop Paw out of the shade of a tree, bush, fence post, even the horses, where he was resting. He was further irritated by young Elwin, a strapping hulk, who crawled out from under his car now and again to shout criticisms of the plowing.

Before this wound had time to heal the Kettle cows started crashing through our fences and eating our fruit trees and our gardens. Beset by flies and long-standing hunger they became a constant menace particularly as the Kettles were experimenting with a small scraggly garden and decided that the quick way to protect it was to mend their own barbed-wire border fences and keep their stock entirely off their property and free to plunder and pillage the entire countryside.

After the cows had broken in for about the tenth time, Bob took them home and stormed into the Kettle yard demanding some immediate action. The dignity and force of his entrance were somewhat impaired by the fact that as he came abreast of the back porch he found himself face to face with Mrs. Kettle who was comfortably seated in the doorless outhouse reading the Sears, Roebuck catalogue and instead of hurriedly retiring in confusion she remained where she was but took active part in the ensuing conversation.

Bob, very embarrassed, turned his back but continued to state his case. "I don't want to quarrel with my neighbors and I know you old people have a hard time keeping up your fences, but by God if your cows don't stay off our place I'll take the car and chase them so damned far into the hills they'll never come home." Maw said, "Why don't you save gas and shoot the bastards?" Paw appeared just then from the cellar where he had no doubt been resting in the shade of the canned fruit, and launched his "The boyth won't HELP ME AND THE OLD LADY and I can't do it all and we fixth the fentheth and THE BUGGERTH GET OUT ANY-WAYS but if you'd come down and give uth a day or two on the fentheth maybe we could KEEP THEM IN . . ." plea, but Maw interrupted with "It's the goddamned bull, Paw, he's did this every summer. Bob, he's et every garden in the valley and he's broke out of every fence and he's got to be shut up."

Paw moved up to lean in the outhouse doorway and said,

"Now, Maw, it ain't the bull, itth the flieth. Perhapth, Bob, if you could give uth a hand with the manure, thay a day or tho, we could get rid of the flieth. . . ." Bob recognized defeat when he saw it and anyway you can't be either threatening or forceful with your back to the audience, so he came home and grimly added a strand of barbed wire to our rail fences and mended the rustic gate.

The cows continued to come and, as summer progressed and the flies got worse, the cows got so they could leap four rails and a strand of barbed wire with the grace and skill of antelopes. Bob became desperate and on advice of other experienced farmers, he loaded his shotgun with rock salt. I doubted at the time that this would do any good since the bull, a wizened sallow little bookkeeper type without a vestige of the lusty manliness which is ordinarily associated with the word bull, quite evidently tried to make up for his lack of physique by telling the cows, "Say girls, if you'll follow me I'll take you to a keen restaurant up on that mountain," and no peppering of rock salt was likely to make him give up his only lure. And I was right. Bob shot and the bull roared and retreated a short distance down the road only to return within the hour to be shot again and to roar and retreat again.

By the end of the first spring Bob hated the Kettles with a deadly loathing and I couldn't blame him—they practically doubled his work and certainly impeded his progress. By the time we had weathered the first winter his attitude had softened somewhat, and by the end of the second year he accepted them like one does a birthmark. I enjoyed the Kettles. They shocked, amused, irritated and comforted me. They were never dull and they were always there.

With misfortune constantly stalking them and poverty and confusion always at hand, I was amazed at the harmony that existed among the Kettles. There was no bickering or blaming each other for things that happened—there was no need to, for the fault didn't lie with them, they figured. Taking great draughts of coffee, Mrs. Kettle told me again and again where the fault lay. "It's them crooks in Washington," she said vehemently. "All the time being bribed and buyin' theirselves big cars with our money." To Mrs. Kettle there was but one Government and that was in Washington, D. C. She had no knowledge of any county, city or state governments. "The whole damn shebang" was in Washington, and Washington to her was a place where everyone was in full evening dress twenty-four hours a day attending balls and dinners which seethed with spies, crooks, liquor, loose women, Strauss waltzes and bribes. Politics were the Kettles' out. When the manure in the barn was piled so high Paw couldn't get in to milk the cows or Tits' Mervin had given her a black eye, or there was no chicken feed or money to buy any, Mrs. Kettle would say, "Look! Just look what them crooks in Washington has did. They put them new fancy laws on time payments so Paw can't get a manure spreader. They give Mervin his Indian money so he gits drunk and hits Tits. They're payin' the farmers not to raise chicken feed and the price is so high I can't git the money to buy it. If you want to know what I think," she would take another strengthening gulp of the coffee, then glaring at Paw, Elwin, Tits and me, would conclude, "I think them politicians can take their crooked laws and their crooked bribes and stuff 'em." They would all nod wisely. The blame had been put squarely where it belonged and nobody on the Kettle farm had to go sneaking around feeling guilty.

The Hickses, our other neighbors, lived five miles down the road in the opposite direction from the Kettles. They had a neat white house, a neat white barn, a neat white chicken house, pig pen and brooder house, all surrounded by a neat white picket fence. At the side of the house was an orchard with all of the tree trunks painted white but aside from these trees there was not a shrub or tree to interfere with the stern discipline the Hicks maintained over their farm. It made me feel that one pine needle carelessly tracked in by me would create a panic. Mrs. Hicks, stiffly starched and immaculate from the moment she arose until she went to bed, looked like she had been left in the washing machine too long, and wore dippy waves low on her forehead and plenty of "rooje" scrubbed into her cheeks.

Mr. Hicks, a large ruddy dullard, walked gingerly through life, being very careful not to get dirt on anything or in any way to irritate Mrs. Hicks, whom he regarded as a cross between Mary Magdalene and the County Agent.

When we first moved to the ranch we were invited to the Hickses to dinner and to an entertainment at the schoolhouse. For dinner we had a huge standing rib roast boiled, boiled potatoes, boiled string beans, boiled corn, boiled peas and carrots, boiled turnips and spinach. Mrs. Hicks also served at the same time as the meat and vegetables, cheese, pickles, preserves, jam, jelly, homemade bread, head cheese, fried clams, cake, gingerbread, pie and tea. This was supper. Dinner had been at eleven in the morning. Mrs. Hicks, a slender creature, ate more than any ten loggers but as she took her third helping she would remark sadly, "Nothing sets good with me. Nothing. Everything I've et tonight will talk back to me tomorrow."

After Mrs. Hicks and I had washed the supper dishes we retired to the tiny living room to sit in a self-conscious circle on the golden oak chairs around the golden oak table and the Rochester lamp while Mr. Hicks fumbled fruitlessly with the radio and Mrs. Hicks firmly snipped off between her teeth any loose threads of conversation. Occasionally she would glance sharply at Mr. Hicks and I felt that one false move and she would take him by the collar and put him outside. After one silence so long that I could feel the tidies of the chair sticking to my neck and arms, Mrs. Hicks called Mr. Hicks into the kitchen and I don't know whether she twisted his ear or what but he announced that he was not going to the entertainment as one of the cows was expecting a calf. Bob elected to stay and help with the delivery and Mrs. Hicks and I set off for the Crossroads in her car. We also shared the car with Mrs. Hicks' liver and her bile, neither of which functioned properly and though she had been to countless doctors and had had several "wonderful goings over" she had to take pills all of the time. She drove, as did all the natives of that country, on the wrong side of the road, very fast and with both hands off the wheel most of the time. During the course of the drive she missed by a hair two other cars, a cow, a drove of horses, a wagon and a road scraper but not a feint in the blow by blow account of the fight between her liver and her bile. Her liver was so sluggish that it had constantly to be primed in order to make it pump her bile, according to Mrs. Hicks. Just before we went into the auditorium of the schoolhouse, she took two of the priming pills and I was very disappointed not to hear liver's motor start and a cheery chug-chug-splash as it pumped Mrs. Hicks' bile into her bilge or wherever bile goes.

During the drive home Mrs. Hicks entertained me with *her* many miscarriages, *her* sisters' many miscarriages, *her* cows' many miscarriages, and *her* chickens' blowouts. The internal structures of Mrs. Hicks and all of *her* connections were evidently so weak that I was relieved when we reached home without the crankcase dropping out of *her* car. When we got in the house, Bob and Mr. Hicks were celebrating the arrival of a heifer calf with a bottle of beer. Mrs. Hicks' disapproval stuck out all over like spines, but when I lit a cigarette she turned pale with horror. "It's not that I mind so much," she told me later, "I know you're from the city but I'd hate to have you smokin' when any of my friends come in because they might think I was the same kind of woman you was."

Mrs. Hicks was good and she worked at it like a profession. Not only by going to church and helping the poor and lonely but by maintaining a careful check on the activities of the entire community. She knew who drank, who smoked and who "laid up" with whom and when and where and she "reported" on people. She told husbands of erring wives and wives of erring husbands and parents of erring children. She collected and distributed her information on her way to and from town, and apparently kept a huge espionage system going full tilt twenty-four hours a day. Having Mrs. Hicks living in the community was akin to having Sherlock Holmes living in the outhouse, and kept everyone watching his step. I was surprised when I learned that Birdie Hicks had a mother —she was so pure I thought perhaps she had come to life out of the housedress section of the Sears, Roebuck catalogue. But one warm evening that spring I left Bob with the egg records and the baby and boldly struck out for Mrs. Hicks' to stitch some curtains on her sewing machine. When I arrived, Mrs. Hicks, her mother and Cousin June were sitting on the front porch slapping at mosquitoes and discussing their miscarriages. After the introductions had been made I sat down for a while before opening my brown paper parcel and exposing the real reason for my visit. This was considered good manners, for in the country where people only call to borrow or return or exchange, and everyone is hungry for companionship, it is considered very impolite to hastily transact your business and leave. You must exchange views of crops and politics if you are a man, gossip if you are a woman, then state your business, then eat no matter what time of day it is, then exchange some more politics or gossip and at last unwillingly tear yourself away. I had sat on Birdie Hicks' front porch for perhaps two minutes when I realized that hungry as I was for companionship this visit was going to be an ordeal, for Birdie's mother, a small sharp-cornered woman with a puff of short gray hair like a gone-to-seed dandelion, tried so hard to be young that conversation with her was out of the question and her ceaseless activity was as nervewracking as watching someone blow up an old balloon. When we were introduced she said, tossing her head about on its little stem, "Bet you thought I was Birdie's sister instead of her mother. Sixty-four years young next Tuesday and everybody guesses me under forty. He, he, he! Everybody does. It's 'cause I'm so active." Whereupon she shot out of her chair and leaped four feet off the ground after a mosquito. Coming down with the astounded mosquito in her little claw, she caught herself deftly on the balls of her feet, bent her knees so that she was almost squatting, then snapped into a standing position, turned and winked at me. I'm not able to wink and nothing else seemed adequate, so I just sat. Cousin June, a plump middle-aged woman, turned to Mrs. Hicks and said, "Honest to gosh, Birdie, she's like a little kid." Mrs. Hicks said rather testily, "For heaven's sake, Ma, set down. You make me nervous." Mother finally perched on the edge of the porch railing but kept her eyes darting, head bobbing and foot tapping and I felt that she had every pore coiled ready for the next spring.

Cousin June laid down her tatting, rolled back her upper lip, exposing enormous red gums sparsely settled with nubbins of teeth, and began an interminable story of a supposingly funny incident that had taken place at the grange meeting. She laughed so much during the telling that it was difficult to understand what she said and either I missed the point or as I suspect there wasn't any because it sounded like "and . . . ha, ha, ha, ha, . . . ho, ho, ho . . . hehehehe . . . owoooooooooo! Well, anyway this fellah say to me . . . ho, ho, ho, ho, ho, hehehehe, hahahahaha, oooooooooooooowh . . . I thought I'd die . . . heheheheh heh . . . hahahahahahah. It's about time you got here . . . hahahahahahah . . . heheheh . . . hohohoho." Mother and Birdie were wiping their eyes and urging her to go on and I felt as left out as though they had all suddenly begun to speak Portuguese. In desperation I began unwrapping my package but this also proved embarrassing as they stopped dead in the middle of a neigh, thinking I had brought Birdie a present. Mumbling apologies I slunk in to sew my seams, but apparently their disappointment was short-lived for above the whirring of the machine I could hear "heheheheheheh, hahahahahahah, this fellah says . . ." "Go on, Junie, what did he say, hahahaha?" "Well, hahahahahahah, hohohohohohoh . . ." and the thuds of Mother leaping about after mosquitoes and being young.

When I had finished my curtains Mrs. Hicks served coffee and heavenly fresh doughnuts and, out of kindness and to explain my stolid dullness, said to Mother and Cousin June, "She reads." Mother in the act of hurling herself at the stove to get the coffee pot, stopped so quickly she almost went headfirst into the oven. "Well," she said, "so you're the one. Birdie's told me all about you and I'm saving my old newspapers for you." I started to say, "Oh, I can't read that well!" but Mr. Hicks came in then and Mother leaped to his shoulders pick-a-back fashion, which evidently delighted him, for his heavy face glowed and he said, "You look younger'n Birdie, Maw. Might be her daughter!" I glanced at Birdie and we felt together that it made no difference how young

Mother looked, for our money, she had lived much too long.

I had meant to leave before it got dark and so didn't bring my flashlight, but the moon was high and the pale green moonlight proved adequate if I discounted stepping high over shadows and coming down with a spine jarring thump into chuckholes. At the top of the second hill a large black bear lumbered slowly across the road just in front of me. He seemed such a pleasant change from Mother and Cousin June that I forgot to be frightened.

The next morning Mrs. Kettle, clad for some mysterious reason in a woolen stocking cap and an old mackintosh, although the day was warm and bright, lumbered up to borrow some sugar. I asked if she knew Mother. She said "Godalmighty yes. Hops around like she was itchy, yellin' 'Don't I look young—took me for Birdie's sister, didn't you?' " Mrs. Kettle's two huge breasts and two huge stomachs plopped and quivered as she imitated the twittering mother. "Always talkin' about how delicate she is. 'Too little to have more'n one kid. Miscarried eight times,' she says. Considerin' the way she jumps around it's a wonder that ain't all she dropped. Acts like a goddamned flea and looks like a goddamned fool!" For that I quickly got the sugar and tossed in a package of raisins.

When it came time to plant the field crops, the potatoes, the mangels, the rutabagas and kale, that second spring, Bob and I decided that rather than work in these plantings between my regular chores we would hire someone and get this work done all at once, and incidentally right. We inquired of the Hickses first about available odd jobbers but they were rather superior about the whole thing and insinuated, and rightly so, that were I more competent Bob wouldn't have to hire help. That Mr. Hicks never had hired anyone in all the twenty years he had had the ranch; that they really wouldn't know whom to suggest. So we tried the Kettles. They, of course, had hired labor. They often took the cream check to pay a man to gather the eggs and haul in feed, which necessitated selling the eggs to buy feed for the cows so they would produce cream to sell, to pay the man, to gather the eggs. This left no money for chicken feed so they would borrow from us as much as they dared and when they didn't dare any more they would let the hired man go, lacking two weeks of his full pay, the chickens would go back to roosting on the front porch and laying in the orchard, the cows would be fed egg mash and the pigs would get the rest of the scratch. The Kettles recommended Peter Moses, a little, old, apple-cheeked man who "odd jobbed" and claimed to be the most patriotic man in the "Yewnited States of America." "Look at them goddamned mountains! Look at them goddamned trees! Look at them goddamned birds! Look at that goddamned water! Every sonofabitchin' thing in this whole goddamned country is purty," he told me with tears in his eyes.

Just before he came to work for us Peter Moses had a job working on the county road. The men were blasting out some stumps so a curve could be eliminated, and it was Peter's job to stand with a red flag and stop the cars before the blast. The mail truck came along and Peter waved it through. "Go on! Goddammit, go on!" he yelled and the mailman drove on and just missed the blast which sent two rocks through his windshield and laid a slab of bark on top of his car. He got out of the truck and walked back. "Hey, Peter, did you tell me to go through?"

"Sure did," said Peter.

"Why, you damn fool," said the mailman. "A blast went off almost under the truck and the rocks broke my windshield. Why didn't you hold me back?"

"Can't do 'er," said Peter. "The Yew S. Mail must go through!"

Mrs. Kettle told us how Peter had appointed himself the official smoker-out of draft dodgers during World War I. Mrs. Kettle said naively, "There was some Germans lived on a ranch up here in the mountains and they had two boys that shoulda went to war but they was hidin' in the hayloft and Peter Moses heard about it and he went up there and seen where they was and reported them to the Government men who had come out to get my boys to enlist." Peter Moses swore that the Kettle boys were so anxious to enlist that they were down in the basement hiding behind the canned fruit, when the Government men came.

The Maddocks had one of the most prosperous farms in that country. Six hundred acres of peat, drained and under cultivation; a herd of eighty-six Guernsey cows; a prize bull; pigs, rabbits, chickens, bees, ducks, turkeys, lambs, fruit, berries, nuts, a brick house, new modern barns and outbuildings; their own water and light systems, and a wonderful garden had the Maddocks. They had also five sons who had graduated from the State Agricultural College and Mrs. Maddock herself was said to be a college graduate. We drove past their beautiful ranch on our way to and from Town and one day there was a sign on the mailbox "Honey for sale." I persuaded Bob to stop. We drove through the gateway and up a long gravelled drive which swept around the house and circled the barnyard. We stopped by the milkhouse and a large hearty man in clean blue-and-white-striped overalls came out, introduced himself as Mr. Maddock and invited us to go over the farm. The farm was everything we had heard. The epitome of self-sufficiency. The cows gave milk to the chickens, the chickens gave manure to the fruit trees, the fruit trees fed the bees, the bees pollenized the fruit trees, and on and on in a beautiful cycle of everything doing its share. The exact opposite of that awful cycle of the Kettles' where Peter robbed Paul to pay George who borrowed from Ed. The Maddock livestock was sleek and well cared for. The barns were like Carnation Milk advertisements–scrubbed and with the latest equipment for lighting, milking, cleaning and feeding; the bunkhouses were clean, comfortable and airy; the pigpens were cement and immaculate; the chicken houses were electric lighted, many windowed, white and clean; the duck pens, beehives, bull pens, calf houses, turkey runs, rabbit hutches, and the milkhouse were new, clean and modern. Then we went to the house. The house had a brick façade and that was all. The rooms were dark–the windows small and few. The kitchen was small and cramped and had a sink the size of a pullman wash basin. In one corner on a plain sawhorse was a wooden washtub. Mrs. Maddock was as dark and dreary as her house, and small wonder. She told me that she hadn't been off the ranch for twenty-seven years; that she had never even been to "Town" or Docktown Bay. When we said good-bye Mr. Maddock shook hands vigorously. "Well," he asked proudly, "what do you think of my ranch?" At last I understood Mrs. Kettle. There was but one suitable answer to give Mr. Maddock and I was too much of a lady.

Mary MacGregor had fiery red, dyed hair, a large dairy ranch and a taste for liquor. Drunker than an owl, she would climb on to her mowing machine, "Tie me on tight, Bill!" she would yell at her hired man. So Bill would tie her on with clothes lines, baling wire and straps, give her the reins and away she'd go, singing at the top of her voice, cutting her oats in semi-circles and happy as a clam. She plowed, disked, harrowed, planted, cultivated and mowed, tied to the seat of the machine and hilariously drunk. A smashing witticism of the farmers was, "You should take a run down the valley and watch Mary sowin' her wild oats."

Birdie Hicks pulled down her mouth and swelled her thin nostrils when she mentioned Mary's name. "She's a bad woman," said Mrs. Hicks, "and we never invite her to our basket socials." I asked Mrs. Kettle about her. She said, "She's kinda hard but she's real good-hearted. There ain't a man in this country but what has borrowed money from Mary and most of 'em has never paid it back. The women don't like her though and all because one time her old man was layin' up with the hired girl and she caught 'em and run a pitchfork into her old man's behind so deep they had to have the doctor come out and cut it out. She said that would teach him and it did because he got lockjaw and died from where the pitchfork stuck him. Mary felt real bad but she said she'd do it again if conditions was the same."

Mary sold cream to the cheese factory. One morning she found a skunk drowned in a ten-gallon can of cream. She lifted the skunk out by the tail and with her other hand she carefully squeezed the cream from his fur. "Just between us skunks, cream is cream," she said as she threw the carcass into the barnyard. She sold the cream and vowed she'd never tell a soul but Bill the hired man told everyone, especially people he saw coming out of the cheese factory with a five-pound round of cheese.

Our first spring on the ranch we didn't have any callers because no one knew we were up there and anyway at that time we didn't have anything to borrow or rather lend nor were we experienced enough to be sought out for advice. Those are the reasons for calling—the time for calling is between four in the morning and seven in the evening and the season is springtime. Summer is too hot, too busy, fall is for harvesting, winter is too wet and rainy. Spring is the time for building, planting, plowing, reproducing and the logical time for calling and borrowing. No one told me this; I learned by bitter experience.

I remember well how the night before I had been awakened by that taut stillness which presages mountain rain. I lay there in the thick dark, at once alert and unreasonably teetering on the edge of terror. No sound, no movement anywhere. Curtains poised in the middle of a sway, half in and half out the window. Shades gone limp. A trailer of my climbing rose clutching the window sill to keep from twitching. Breezes on tiptoe. Trees reaching. Trees bent listening. Everything in the mountains playing statue. Then the signal. Tap, tap. Tap, tap, tap. Tap, tap, tap, tap, tap, tap, tap. A great, soft sigh spread through the orchard, across the burn, over the mountains, everywhere. A frog croaked, the curtains bellied, a shade rattled, an owl hooted apologetically and the rain settled down to a steady hum.

I got up the next morning to a dreary world of bone-chilling air, wet kindling, sulky stove and a huddled miserable landscape. It was Spring's way of warning us not to take her for granted.

It took me from four o'clock until seven-thirty to care for my chicks, get Stove awake and breakfast cooking. Each time I went outdoors I was soaked to the skin by the rain, which was soft, feathery and scented but as penetrating as a fire hose. After using up three sets of outside garments, in chilly desperation I put on my flannel pajamas, woolly slippers and bathrobe until after breakfast. What luxury to be shuffling around in my nightclothes getting breakfast after all those months of being in full swing by 4:15 A.M. with breakfast a very much to the point interval at five or five-thirty. When Bob came in he acted a little as if he had surprised me buttering the toast stark naked. I patiently explained the reason for my attire and was defiantly pouring the coffee when a car drove into the yard. "Dear God, not callers at 7:30 and on this of all mornings!" I prayed. But it was.

A West-side dairy rancher and his sharp-eyed wife. Mr. and Mrs. Wiggins. Mr. Wiggins wanted some advice on fattening fryers and she wanted to look me over. It was very natural on her part as she had probably heard from Birdie Hicks that I smoked and read books and was a terrible manager, but she didn't have to sit on a straight chair in the draughtiest corner of the kitchen with her skirts pulled around her as though she were waiting for her husband in the reception room of a bad house.

I implored Bob, with every known signal, not to leave me alone with this one man board of investigation, but Bob went native the minute he saw another rancher and became a big, spitting bossy *man* and I was jerked from my pleasant position of wife and equal and tossed down into that dull group known as *womenfolk*. So, of course, Mrs. Wiggins and I were left alone. I tried to sidle into the bedroom and slip on a housedress and whisk everything to rights before the baby awoke, but the puppy chose that moment to be sick and instead of throwing up in one place he became hysterical and ran around and around the kitchen belching forth at intervals and mostly in the vicinity of sharp-eyed Mrs. Wiggins. She pulled her feet up to the top rung of her chair and said, "I've never liked dogs." I could see her point all right but it didn't improve the situation any, especially as Sport, our large Chesapeake retriever, managed to squeeze past me when I opened the back door to put the mop bucket out, and bounded in to lay first one and then the other large muddy paw on Mrs. Wiggins' starched lap. She screamed as though he had amputated her leg at the hip, which of course waked the baby. I retrieved Sport and wedged him firmly in behind the stove, we exchanged reproachful looks, I wiped up his many many dirty tracks, sponged off Mrs. Wiggins and picked up small Anne. As I bathed the baby, Mrs. Wiggins handed me flat knife-edged statements, as though she were dealing cards, on how by seven o'clock that morning she had fed and cared for her chickens, milked five cows, strained and separated the milk, cleaned out the milkhouse, cooked the breakfast, set the bread, folded down the ironing and baked a cake. It took all of the self-control I had to keep from screaming, "SO WHAT!"

Mrs. Wiggins, no doubt, had quite a juicy morsel for the next basket social, but I had learned my lesson and from that day forward I was ready for Eleanor Roosevelt at four-seven in the morning.

9

I Learn to Hate Even Baby Chickens

PRIOR to life with Bob my sole contact with baby chickens had been at the age of eleven. Lying on my stomach in our hammock which was swung between two Gravenstein apple trees in the orchard by the house in Laurelhurst, I pulled out grass stems, ate the tender white part and watched Layette, Gammy's favorite Barred Rock hen, herd her fourteen home-hatched fluffy yellow chicks through the drifting apple blossoms and under the low flowering quince trees. This sentimental fragment of my childhood was a far cry from the hundreds and hundreds of yellowish white, yeeping, smelly little nuisances which made my life a nightmare in the spring.

I confess I could hardly wait for our chicks to come and spent many happy anticipatory hours checking the thermometer and reveling in the warmth and cleanliness of the new brooder house. But I learned to my sorrow that baby chickens are stupid; they smell; they have to be fed, watered and looked at, at least every three hours. Their sole idea in life is to jam themselves under the brooder and get killed; stuff their little boneheads so far into their drinking fountains they drown; drink cold water and die; get B.W.D.; coccidiosis or some other disease which means sudden death. The horrid little things pick out each other's eyes and peck each other's feet until they are bloody stumps.

My chick manual, speaking from the fence said, "Some chicks have a strong tendency to pick and some don't." (I was reminded of the mushroom book's, "Some are poisonous and some are not.") The chick manual went on to say, "The causes of picking are overcrowding, lack of ventilation or cannibalism." Our chicks, according to the standards set by the manual, had plenty of air and space so I added plain meanness to their list of loathsome traits. From the time of their contemplation, our baby chickens were given the utmost in care and consideration and their idea of appreciation was to see how many of them could turn out to be cockerels and how high they could get the percentage of deaths. I knew that Layette's babies never acted like that, which was a flaw-proof argument for environment over heredity and against any form of regimentation.

I really did my badly organized best to follow my chicken manual to the letter, even though it required that I spend one out of every three hours in the brooder house—measuring feed, washing water fountains, removing the bloody and the dying to the first-aid corner—and all of my leisure time nailing a dead chicken to a shingle, splitting the carcass from stem to stern and by peering alternately inside the chicken and at a very complicated chart, trying to figure out what in the world it died of. I always drew a blank. In my little Death and Food Record book, I, in my prankish way, wrote opposite the date and number of deaths, "Chickenpox-Eggzema and Suicide." When he checked the records, Bob noted this fun-in-our-work, and unsmilingly erased it and neatly wrote, "Not determined." Men are quite humorless about their own businesses.

My chick manual was detailed to the extent that it gave the number of minutes it should take so many chicks to clean up so much food—what to feed every single day until the chickens were six weeks old; even what to do about the floors, hovers, founts, hoppers, etc., four weeks before brooding. From my experience I would supplement this prior-to-brooding advice to read, "Four weeks before brooding, leave on an extended trip to the Baranof Islands."

I well remember how the Lucrezia Borgia in me boiled to the surface as I read in my chick manual, "A single drink of cold water may be fatal to a baby chick." "You don't say," I thought, licking my fevered lips and glancing longingly at the little lake filled with icy water. But my poultricidal tendencies were replaced with pure hysteria as I read on, "Water may be warm when you put it in the founts, but will it stay warm?"

"My God, isn't it enough that my hands will soon be dragging on the ground from carrying buckets and buckets and buckets of water, and that Stove has acquired a permanent list on his reservoir side, without being further tortured with trick questions? Why don't you get underneath the brooder and see if the water stays warm, you big bore? Me, I'll fill the fountains with warm water and curses every three hours and take a chance." That was my reaction to my chick manual.

The next cozy paragraph was headed "Dopey Chicks." "If many chicks are 'dopey' and you are sure they are not overheated or gassed, those chicks and the chicks that continually chirp should be sent to the nearest pathological laboratory (to see who's dopey?). If the report says B.W.D., it is better to disinfect the premises and start new chicks." I could find no explanation of B.W.D., but to me it was code for the best news in the world. It might have been better to start new chicks, but it might have been best to take the next train for Mexico.

I wondered how other chicken ranchers' wives reacted to

baby chickens. Was there something in my background which kept me from becoming properly adjusted to the chicken, or was there just that too wide a gulf separating a woman and a chicken? I was delighted therefore, one spring morning, to have Mrs. Hicks halloo from the road and invite me to ride down to Mrs. Kettle's with her while she returned some bread pans. Both Mrs. Kettle and Mrs. Hicks were raising baby chickens and I thought this would be a splendid opportunity to make comparisons and to slip out of harness for a little while.

I had bathed and fed small Anne and put her to sleep in her carriage in the orchard, so I took a quick look at all of my other babies to be sure they were well fed and asleep, threw Bob a few hazy instructions, hung my apron on the gatepost, and we were off. Mrs. Hicks, full to the lip with some new and wonderful bile primer, was cheerful to the point of gaiety. Not so Mrs. Kettle, who clumped morosely out to greet us, kicking at her beloved mongrels as she went by.

At first I thought it the heavy curtain of gloom which made the spacious kitchen seem so crowded—then I became conscious of a rising crescendo of twitterings from the vicinity of the stove. Mrs. Kettle was rearing her baby chickens in the kitchen. That area back of the large woodstove which ordinarily housed the woodbox, the house slippers and barn boots of Mr. Kettle and the boys, a couple of bicycles, bits of harness, the newspapers, the dogs and cats and the car parts, had been turned into a brooder house. Fenced off by rusty window screens leaning against chairs and heated by a varied assortment of jars, cans and bottles filled with hot water, two hundred baby chicks existed in apparent health and contentment. No B.W.D. there. No disinfectant, no thermometer—and no sickness either. "That manual writer should see this," I thought bitterly.

Mrs. Kettle was also harboring in her kitchen a little runt pig, the sole survivor of a litter eaten by its mother. "The old bitch ate 'em all but this little bastard," chronicled Mrs. Kettle, whose nomenclature was always colorful but at times confused.

The chicks she dismissed lightly with "Paw ordered 'em last fall but didn't git around to buildin' the brooder house before they come so I guess we'll just have to raise 'em in here." The chirping chickens and the little pig clicking around under foot on his little sharp hoofs, all completely innocent of any form of housebreaking, didn't bother Mrs. Kettle a whit. What did trouble her was the fact that her elder sister, who twenty-odd years before had had the good fortune to marry a man both wealthy and prominent, had had the effrontery to send Mrs. Kettle by the morning mail, in lieu of a rich gift, an enormous tinted portrait of herself in evening dress. This Mrs. Kettle had set up on the table, easeled by the cracked white sugar bowl and a jar of jam.

Scratching herself vigorously and gesticulating with her soup ladle, she sneered, "Look at that, would you—pretty fine ain't we with our dinners all bare like a whore's?" (The dress was cut in a very modest V.) "And covered with jools which your old man got from bribing the Government. Well, you can stuff your jools and your crooked husband and—" Mrs. Kettle's face brightened. "You know where I'm going to hang your goddamned pitchur? In the outhouse!"

Mrs. Hicks and I took our leave at this point, but as we drove over the hill we heard the sound of violent pounding as Mrs. Kettle hung sister's gilt-framed picture.

Mrs. Hicks invited me to go home with her for a cup of coffee and to see her baby chickens. I accepted instantly, of course, so we jounced right past our ranch and down the mountain on the other side.

The coffee, strong and delicious, with thick yellow cream, was accompanied by that heavenly and completely indigestible delicacy, fried bread. Apparently all Mrs. Hicks did was to drop twisted pieces of bread dough into hot fat and in a minute or two take out big golden brown puffs which she dipped in powdered sugar and covered with strawberry jam. They weren't small and had what I'll call body, but I ate three and Mrs. Hicks five before we made a move toward the chicken houses. Then I tried a sprightly leap off the back porch, only to find that I had suddenly been outfitted with ballbearings. The fried bread rolled from side to side giving me the feeling of sea legs. I glanced at Mrs. Hicks but she sailed ahead of me like a piece of thistledown. Thistledown or no, I already had a different conception of her liver and vowed that in the future I would be a little more careful of what was left of mine.

Mrs. Hicks' brooder house smelled so strongly of disinfectant it made my eyes water and the chickens, looking as if they had sprouted under boards, drooped listlessly around the edges of their immaculate modern house. Gammy used to say, "Too much scrubbing takes the life right out of things," but a perennial droop seemed to be Mrs. Hicks' yardstick of cleanliness.

On the ride home I clutched my fried bread on the rough places and shifted it left and right on the curves, while Mrs. Hicks, seemingly in perfect comfort, chatted gaily. I asked her about the percentage of deaths in her chicks and was amazed to learn that out of five hundred chicks she had lost only five. She said, "Those five died the day after we got the chicks and I don't think they was right, but just in case it was anything catching I put a little disinfectant in the drinking water and the rest pulled through fine." What I think really happened was that Mrs. Hicks called a meeting of her chicks right after they arrived and told them, "I'm the boss here and I'm not going to put up with any sickening or dying. The first chick I catch dying is going to get what for and I mean it." And the chicks, disinfected inside and out, stayed alive—or else.

Mrs. Hicks was really a remarkable woman. She was slender and frail-looking, but she did so much work that just to hear her tell about it made me tired. She took all of the care of the chickens, the calves, pigs, turkeys, ducks and eggs, in addition to keeping her house like an operating room, baking, cooking, cleaning, sewing, washing and ironing. In winter Mr. Hicks, as did most of the farmers, supplemented his income by longshoring at Docktown or working in the lumber camps. During these times Mrs. Hicks did all of her usual work and milked ten cows night and morning, separated the milk, fed and watered the horses and still had time to take the eggs to town and pick up her spy reports.

Often after a particularly gruelling day, as I banged my shins against the oven door and cursed the inadequacy of coal-oil lamps, I would think enviously of Mrs. Hicks, who at that moment was probably standing in her immaculate kitchen, in an immaculate apron and housedress, wondering, now that the dishes were done, if she shouldn't just bake an angel food cake or set some rolls for the basket social. Just thinking of her in her tireless efficiency sometimes made me think I had better give up smoking and take up bile priming in its stead.

Once Mr. Hicks got hurt in the woods and was sent to a hospital in town, Mrs. Hicks went in to stay with him and Bob and I took care of their ranch for them for a few days. I couldn't begin to take over all of Mrs. Hicks' duties, but between us we managed very well, except that I fixed the milk and cream which we bought from the Hickses and I evidently used the wrong faucet on the separator because the cream, instead of being the top-milk variety which we had been getting all spring and summer, oozed into the bottle, dark yellow and thick. I didn't say anything to Bob, for he leaned terribly toward fair play and would probably have left no stone unturned until he had located the error, but I noticed that for those few days he used cream on everything but his meat. Every day I unlocked Mrs. Hicks' back door and tiptoed into the house and dusted the golden oak furniture and resisted a strong impulse to rummage in her bureau drawers and pantry—a holdover from the days when I was a child-sitter and supplemented my 25c-for-the-afternoon pay by eating everything not nailed down in the houses of my customers.

Working within the sacred bounds of Mrs. Hicks' cleanliness proved such a strong impetus for a while that I found myself going after corners in my own house with pins and washing the face of the kitchen clock. I waited for her return with the smug feeling of someone who has done something well and knows he is going to be praised. Mrs. Hicks was very grateful to Bob and me and she and Mr. Hicks told us over and over what kind neighbors we were, but the next day Bob and I stopped on our way to town to see if we could get them anything, and Mrs. Hicks had her washtub filled with boiling water and disinfectant and soapsuds and was scrubbing the walls and floors of the chicken houses, calf houses, pig houses, turkey houses, duck houses and brooder house which Bob and I thought we had kept so clean. I gave up.

Bob turned out to be the best chicken farmer in our community. He was scientific, he was thorough, and he wasn't hampered by a lot of traditions or old wives' tales. Bob didn't believe in mixing breeding and egg raising—he said that they were separate industries and should be treated as such. His theory was that an egg-raising flock should be kept to a 90-96 per cent lay as much of the year as possible, but that if you were also using the flock for breeding and hatching eggs, such a strenuous laying program weakened the stock and made for poor chicks. He evidently knew what he was doing for his chickens laid eggs and didn't get sick and we always made money. Bob said that he could make money if eggs dropped to 15c a dozen. They never did—I think that 19c was the lowest we ever got and that was in the spring when eggs were plentiful—but Bob was not one to make promises he couldn't keep. Bob said that the secret of success in the chicken business for one man was to keep the operation to a size that could be handled by one man. He estimated that one man could handle 1500 chickens (provided his wife was part Percheron) by himself and make a comfortable living—but most people's trouble was that they were so comfortable on 1500 chickens that they figured they might as well be luxurious and have 2500. Then the trouble started: they had to hire help; they had to have much more extensive buildings and equipment; and to warrant the extra expense they would have to have five or ten thousand chickens instead of 2500. It sounded reasonable, and if Bob said it, it probably was.

An average white Leghorn hen laid from 150 to 220 eggs a year. She cost from $2.25 to $2.50 to raise—this included cost of equipment and bird. Eggs averaged over the year 31c a dozen. Using this as a basis we figured that a hen the first year might, if she tried, lay 204 eggs or 17 dozen, which at 31c would be $5.27. Less her original cost of $2.35, less feed costs of around $2.40, this would leave a profit of about 50c per hen the first year. The second year the eggs were all profit except the feed, unless you wanted to split the cost of the new pullets and bring down the original cost per hen. There was a prize flock of 455 pullets in that vicinity which laid 243.5 eggs per hen per year, 111,027 eggs per year per flock—and made a profit of $3.46 per fowl above feed costs. Our records showed that we were not too far behind this prize flock the second year and we had 1000 chickens.

I kept all of the egg records. I wrote on a large calendar in the kitchen the number of eggs we gathered at each gathering. At the end of the day these figures were entered in a daybook and later entered in a weekly column, along with the feed, which was delivered once a week. It was a very simple system, but when it came time to draw weekly and monthly percentages I was apt to find the hens in the throes of a 150 per cent lay, and then I would have to go laboriously back and try to find out how far back and in which branch of my arithmetic, adding, multiplication or subtraction, the trouble lay.

The percentage of cockerels was a vital factor in determining the cost of each pullet, and I watched the baby chicks with beating heart for the first signs of the little combs which would tell me how we stood. As soon as we could tell them apart, we separated the cockerels and put them in fattening pens where they ate and fought and crowed until it was time to dress them for market. Anything else that I had cared for from birth would have become so embedded in my feelings I would have had to gouge it out, but I got so I actually enjoyed watching Bob stick his killing knife deep into the palates of fifty cockerels and hang them up to bleed. My only feeling was pride to see how firm and fat they were as we dressed them for market.

I got so I could dress chickens like an expert, but have wondered since how this ability to defeather a chicken in about two minutes without once tearing the skin, my only accomplishment, could ever be mentioned socially along with swimming and diving, or gracefully demonstrated as with violin and piano playing. Wouldn't you know that I would excel in chicken picking?

About the time the cockerels were ready for market, the pullets were ready to be taught to roost in their own little houses instead of in the trees, where they were easy prey for owls and wildcats. This meant that at dusk each night Bob and I had to go through the orchard plucking squawking, flapping birds out of the tops of the trees, holding them by the ankles with heads down. When we had as large a bouquet as we could hold, we took them to the pullet houses and planted them firmly on the roosts. At first I felt like a falconer and found the work rather exhilarating, but after about two weeks, when there was still a large group of boneheads who preferred to sleep out of doors and get killed, I found myself inclining toward the you've-made-your-bed-now-lie-in-it attitude.

Chickens are so dumb. Any other living thing which you fed 365 days in the year would get to know and perhaps to love you. Not the chicken. Every time I opened the chicken house door, SQUAWK, SQUAWK-SQUAAAAAAAAWK! And the dumbbells would fly up in the air and run around and bang into each other. Bob was a little more successful—but only a little more so and only because chickens didn't bother him or he didn't yell and jump when they did.

That second spring Bob built a large new yard for the big chickens—the old one was to be plowed and planted to clover, which disinfected the ground and provided greens for the hens. We eventually had four such yards so that by rotation our hens were always in a clean green playground. Other chicken ranchers shook their heads over this foolish waste of time and ground. They also scoffed at feeding the chickens buttermilk and greens the year round. They had been brought up to believe that women had tumors, babies had fits and chickens had croup; green food and fresh air were things to be avoided and a small dirty yard was all a chicken deserved.

Bob paid no attention to the other farmers, and when the new yard was finished we lifted the small runway doors and watched the hens come crowding out, scolding, quarreling, singing, squawking, choosing their favorite places and hurrying like mad to enjoy their playtime. They were gleaming white with health and spring, and didn't seem nearly so repulsive as usual.

When the pullets began laying, Bob and I culled the old hens. We did this at night. We'd lift an old hen off the roost, look at her head, the color of her comb, her shape, her legs, and if we were in doubt we'd measure the distance between her pelvic bones—two fingers was a good layer. Chickens could be culled in the yard except for the trouble encountered in catching them. The good layers looked motherly, their combs were full and bright red, their eyes large, beaks broad and short, and their bodies were well rounded, broad-hipped and built close to the ground. They were also the diligent scratchers and eaters and their voices seemed a little lower with overtones of lullaby. The non-producers, the childless parasites, were just as typical. Their combs were small and pale, eyes small, beaks sharp and pointed, legs long, hips narrow, and they spent all of their time gossiping, starting fights, and going into screaming hysterics over nothing. The non-producers also seemed subject to many forms of female trouble—enlarged liver, wire worms, and blowouts (prolapse of the oviduct). What a bitter thing for them that, unlike their human counterparts, their only operation was one performed with an axe on the neck.

I really tried to like chickens. But I couldn't get close to the hen either physically or spiritually, and by the end of the second spring I hated everything about the chicken but the egg.

I especially hated cleaning the chicken house, which Bob always chose to do on ideal washing days or in perfect gardening weather. In fact, on a chicken ranch there never dawns a beautiful day that isn't immediately spoiled by some great big backbreaking task.

Our chicken house was very large and was complicated with rafters and ells and wings. Cleaning it meant first scrubbing off the dropping boards (which were scraped and limed daily) with boiling water and lye; then raking out all the straw and scraping at least a good half inch from the hard dirt floors; then with a small brush—a very small brush—I brushed whitewash into all the cracks on the walls, while Bob sprayed the ceiling. Then Bob sprayed the walls and criticized my work on the crevices (the only thing he failed to make me do was to catch the lice individually); then we put clean straw all over the floor; filled the mash hoppers; washed and filled the water jugs and at last turned in the hens, who came surging in filled with lice, droppings and, we hoped, eggs.

10

The Lure of the Tropics

I THINK seed catalogues are the most exciting things there are. And I think seed companies are the most generous, for they never question your motives when you write for their catalogues. By looking at the return address in the upper left-hand corner of the envelope they could have seen that I lived in the vicinity of the "most westerly tip of the United States," and yet they never hedged about sending me gorgeously illustrated catalogues mostly devoted to tropical plants with thrilling pictures of orange trees in full fruit and bloom, lemon trees, magnolias, avocados, peppers and other brilliantly colored warm-sounding names like Canna, Iberian Fire Lily, Mexican Flame Flower, African Daisy. On gray soggy November days I pored over last year's catalogues, and after an hour or two I could look out at the squishy landscape without shivering, for I could almost hear the hum of bees, feel the summer heat and see the yard wallowing in tropical glory.

When the new catalogues came in the spring I devoured them and with pencil and paper made lists, which usually totaled around $279 and had to be slashed and slashed. At last I ordered my seeds and spent days rigid with expectancy. I always bought against my better judgment some of the flame-fire-veldt type of plants from a little known semitropical seed company, which invariably substituted Nasturtiums for Belgian Congo Moon Glow Blooms ("often attaining a size of two feet in diameter")—California Poppies for East Indian Pompoms and never put in more than three seeds to a package. They were not very honest, but I could warm my hands over the pictures in their catalogues.

Bob, who had already ordered and received all of his seeds weeks before from a well-known local firm, listened resignedly to my feverish accounts of the front yard exploding with giant Cannas, the house crawling with Flame Flowers, gourds and monstrous Congo Roses, the fences completely hidden by "Unusual Annuals"; then dug a trench the full length of the vegetable garden, filled it with chicken manure, rich brown earth and sweet peas. I saw defeat coming as relentlessly as old age.

The second spring Bob made me a coldframe all my own for my tropical galaxy of bloom. It was not exactly a spontaneous idea on Bob's part, and he was pretty tightlipped as he grimly slammed the nails home, mumbling about the drop-of-water-on-stone technique; but it was a beauty, facing south, with three sections, sashes on hinges and little arms to brace the sashes when I wanted to lift them. When it was finished all I had to do was to wait for my seeds to come and for the earth to take a slight detour from its ordinary whirl on its axis, and success would be mine.

Meanwhile it was time to work on the vegetable garden. As we plowed and harrowed and dragged the feathery loam, I thought of New England people and of how they have to build their soil out of humus and sweat, and it made me feel guilty. Our soil was so wonderful that I could thrust my arm clear to the shoulder in it when it was ready to plant. It was a natural sandy loam and, with chicken manure and compost added, it was so fertile it was almost indecent. Bob made garden rows as straight as dies, spaced to the inch, and his seeds came up the correct distance one from another. When he planted a seed it immediately got busy and sprouted and appeared in exactly the allotted time. From thence forward the progress was about as fascinating and as trustworthy as a Postal Savings Bond. The seed reached full maturity, with interest, in exactly the promised number of days, and another good investment was harvested.

Bob's garden was a thing of symmetry and beauty. It was bordered with great clumps of rhubarb that had bright-red speckled stalks as big as my arm and so crisp they snapped in two, drippy with juice. Between the rhubarb plants, and nourished by the continual waterings of manure water which we gave the rhubarb, were parsley, chives, basil, thyme, sage, marjoram, anise and dill. I put parsley in everything but ice cream, Bob said, but even he admitted that tomato sauce, stew, kidney sauté, spaghetti, or meat pie seemed tasteless without fresh basil once you had tasted them cooked with it. Mint grew in a thick hedge by the woodshed and my great fear was that it might get started somewhere else, it was so eager. In rows about fifty feet long, stretching from the sweet peas to the rhubarb and herbs, were peas, early and late, carrots, turnips, beets, salsify, celery, celery root, lettuce, endive, broccoli, cabbage, cauliflower, Swiss chard, sweet corn, parsnips, beans, cucumbers, tomatoes, squash, radishes, onions (the sweet flat Bermudas which grew as large as apples and almost as mild), and Brussels sprouts. Over by the brooder house—at least that year's location of the brooder house—Bob made an asparagus bed, which I estimated, when in full production, would take care of that portion of the United States extending from the Columbia River to the Pacific Ocean. Our growing season was short, or rather our maturing season was short, for we had late frosts and little hot weather; but the mild winters and long cool springs seemed to give a succulency to the vegetables which I have never seen surpassed. Nothing was ever pithy or tough or harsh flavored, and even carrots left in the ground all winter (by mistake) were crisp and tender the next spring. That was truly a gardener's paradise except that we couldn't grow lima beans or egg plant or melons or peppers and we had to mulch our English walnut trees and apricot trees very heavily, and even then the blossoms were often frostbitten.

With Bob's gardening so stable—so trustworthy—it didn't seem fair to me that mine should always be the wildcat variety. A great many of my seeds not only did not sprout but disappeared entirely. The others came up the next morning like Jack's Beanstalk or didn't show up until I had given up hope and planted something else on top of them. And then they all appeared together and created confusion and ill will. My seeds, no matter in what order they were put into the ground, always came up in bunches. A big clump here, nothing at all here for quite a way, now a little wizened group, a single plant and another huge clump. Also my plants were usually not healthy, and I'm sure that I have introduced more varieties of plant disease than anyone in the Northern Hemisphere. I planted nasturtiums and bachelor's-buttons and they came up covered with South African Jungle rot and Himalayan Spot Wart. I decided at last that instead of the green thumb I had the touch of death and that I was never destined to be a second Mowgli or Little Shepherd of the Hills because I hated little wild things and they hated me. Bob, on the other hand, had nature by the scruff of the neck and his sweet peas had blooms like gladiolus and stems about ten feet long.

The man from whom we bought our bulbs gave me all of

his lovely single dahlia tubers and kept only the hydrocephaloid monsters in liverish lavender and virulent pink. He said, "You ought to get you another hobby, there is some folks who just don't have the feeling. Yep, you should get you another hobby."

11

The Mountain to Mohammed

THAT FIRST YEAR no one, or very few, knew that we were up there in the mountains on our ranch and so we were skipped by the door-to-door sellers. I didn't learn about that delight of country living until one drear day late the first fall. Bob was out in the woods usefully and gainfully employed cutting shingle bolts and I was rattling around in the house longing for my lovely big noisy family and hating the mountains, when a little black truck sidled into the yard, a small man alighted and crept to the back door, where he scratched like a little mouse. I rushed to the door and he was so heartened by my greeting, not knowing that I was glad to see anybody, that he hurriedly scrambled back into the truck and came staggering back under four great black suitcases. He opened the first one and I realized that at last I was face to face with the creator of the knitting book outfits. The coat sweater made like a long tube with an immense shawl collar. The tatted evening dress. The lumpy crocheted bed-jacket tied with thousands of little ribbons. The great big tam. The slipover sweater with the waistline either crouching in the armpits or languishing just above the knees. Jack the Knitter had them all and mostly in maroon, a pink so bright it could have given a coat of tan, and orchid.

Jack the Knitter was so proud of his wares that I had a feeling he made them himself, and I pictured him stopping his truck halfway down the mountain and crawling into the back to tat up another evening dress or maybe knit another big purple tam.

He did have lovely suits knitted of yarn as soft and light as kitten's fur and seemingly devoid of furbelows. I tried one on and was bitter to find the V-neck crowding the crotch for position and the skirt so tight that when I stood erect it made me look as if I were using my fanny for antennae and feeling around for a seat.

As I turned down offering after offering, Jack's little watery eyes grew sadder and his little black suit shinier and shabbier, and we were both pretty desperate when he brought out the socks—the really superior wool socks. I immediately ordered twelve pairs for Bob and begged Jack to stay for dinner. He, however, seemed to have revived considerably under the stimulus of the order and briskly declined, packed his wares and left in the direction of Mrs. Kettle's, where he undoubtedly had dinner after selling all of his sweaters including the expectant-sitter suits.

Bob returned from the woods and pointed out that I had ordered the twelve pairs of socks two sizes too small. I tried to make up my mind whether to walk in the dark and rain down to Mrs. Kettle's and tell Jack the right size or to let the wrong ones come and go through the long exchanging process. I decided to let the wrong ones come and went shuffling off to bed with my will power dragging around my feet like let-down suspenders.

Then came Christmas and the mail order. I bought all my Christmas presents from the catalogues, which was wonderful for I had the thrill of first choosing what I would buy if I could have anything, and then what I would actually buy.

My greatest hurdle in ordering was whether or not to take

a chance on "multi-colored." Practically everything except farm machinery came in pink, blue, green, maize (never yellow) and multi-colored. It sounded as if they had stirred all of the left-over colors together as we used to do with Easter egg dyes. I was strong and limited myself to plain colors, but I felt I had erred. What if I died and Bob sent for and buried me in a multi-colored shroud and I had never seen one. I was also fascinated by that wonderfully ambiguous term "floral background." Since a flower could be anything from a forget-me-not to a Yucca, that was certainly buying a pig in a poke; and so I bought many "floral backgrounds" for my family.

Then came spring and the Stove Man. Along in January, Stove developed virulent digestive trouble. In fact, where his grate had been there was a gaping hole and I had to build my fire like a blazing fringe around the edge. Stove was "taken" in January, but it was March before anything was done about it. We of the mountains didn't dash into town for a new grate. We wore our already taut nerves to within a hair of the snapping point trying to cook on the circle of fire or that faint warm draught that wafted from the ash pit down around Stove's feet, where the pieces of grate and all of the wood had fallen, and waited for a mythical character known as the Stove Man who was supposed to make rounds in the spring.

One morning I had reached the stage where I was craftily planning to chop up a chair or two and build a fire in the sink in an effort to drive Bob to some immediate action, when the Stove Man arrived. With him also were a truckload of stove parts and tools, his wife and three-year-old daughter. Stove Man quickly disembowled Stove, something which I had been longing to do to that big black stinker all winter, spread the entrails all over the kitchen floor and went out to the chicken house to point out to Bob the many opportunities for failure in the chicken business.

This black-future attitude was not from any manic depressive tendencies on the part of Stove Man, but was the reflected attitude of the farmer. The farmers wanted to be sad and they wanted everyone who called on them to be sad. If your neighbor's chickens were each laying a double-yolked egg every single day, all of his cows had just had heifer calves, his mortgage was all paid, his wheat was producing a bushel per stalk and he had just discovered an oil gusher on the north forty, you did not mention any of these gladsome happenings. Instead, when you looked at the chickens, you said, "A heavy lay makes hens weak and liable to disease." The neighbor, kicking sulkily at the feed trough, would reply, "Brings the price of eggs down too."

When you went into the barn you looked over the heifer calves and said, "Lots of t.b. in the valley this year. Some herds as high as 50 per cent." The neighbor said, "Contagious abortion is around too." Leaning morbidly on the fence around the groaning wheat fields, you said, "A cloudburst could do a lot of damage here." The neighbor said, "A heavy rain in harvest time would ruin me." I learned that our farmers were like those women who get some sort of inverted enjoyment out of deprecating their own accomplishments—women who say "This cake turned out just terribly!" and then hand you a piece of angel food so light you have to hold it down to take a bite.

Our farmers were big saddos and our farmers' wives were delicate. Farmers' wives who had the strength, endurance and energy of locomotives and the appetites of dinosaurs were, according to them, so delicate that if you accidentally brushed against them they would turn brown like gardenias. They always felt poorly, took gallons of patent medicines and without exception they, and all of their progeny, were so tiny at birth that they slept in a cigar box and wore a wedding ring for a bracelet.

Stove Man's wife—"Just call me Myrtle"—and Darleen, the small white thready-limbed child, stayed in the house while Stove Man and Bob made their gloomy journey around the ranch. As lunchtime approached with Stove still in surgery, Myrtle and I, accompanied by the drop, drop, drop of her leaky heart, made sandwiches and brewed tea over some canned heat. I offered to warm some soup or vegetables for Darleen, but Myrtle demured with "That kid eats anything —strong as a horse."

The horse ate for her lunch one white soda cracker and a sweet pickle; then hung on the back of her mother's chair and whined. Not bothering to turn around and not missing a mouthful, Myrtle comforted her with threats of "I'll warm your bottom"; "I'll turn you over to your Dad"; "I'll lock you in the truck"; "I'll send for the bogey man"—all of which Darleen ignored and kept on swinging and whining. In desperation, I suggested a nap, but Myrtle said, "That kid has never took a nap since she was weaned—just don't need sleep —strong as a horse." Even though Darleen looked like something they had whipped up out of pipe cleaners, she certainly had endurance. She whined and swung until after five, when they left.

Much to my surprise Mr. Myrtle became very businesslike after lunch and put Stove back together with new grates and a new, more thorough ash shaker. Unfortunately Myrtle also became businesslike and insisted on cutting out dresses for small Anne from the eight lengths of dimity and nainsook I had brought from town the day before. I had patterns but she scoffed at these as totally unnecessary for children's clothes.

When I came in from one of my various trips to the chicken

house, baby buggy, feed room or coldframe, I found Myrtle slashing out the last dress. She stacked the cut-out garments, took Darleen to the outhouse and, with "See you next year if my heart holds out"—"Quit that, Darleen, or I'll smack you!" —"Hope the stove holds together," they were gone.

After dinner that night I examined the ready-to-sew dresses. Something seemed to be wrong. There were eight large round pieces of material—one from each length—and eight rims from which the circles were cut. I was unable to determine which were the dresses and which the scraps. I laid small obliging Anne on the bed and tried to fit her into these strange pieces, but all we were able to work out were the foundations for eight old-fashioned sweeping caps and matching ruffles for Anne's fat posterior.

Either Myrtle had a splendid idea which I was too stupid to grasp or Anne was the wrong shape. "Oh, well, see you next year, Myrtle," I said, dusting off the bedside table with one of the circles and putting the rest away in my bottom drawer, where they remained as long as we lived on the ranch.

For several weeks after the visit of the Stove Man the weather was clear and bright and we worked like maniacs to get caught up with Spring, which raced ahead of us each day, unfolding new tasks for us to do and cautioning us about leaving the old ones too long. Each night Bob drove the truck down into a small valley below the house and filled ten, ten-gallon milk cans with water, and the next morning as soon as I was dressed I filled Stove's reservoir and my wash boiler, and between chores I washed all day long. The clothes billowed and flapped whitely against the delphinium blue sky and the black green hills, and at night I brought in armloads of clean clothes smelling of blossoms and breezes and *dry*.

For three weeks I washed all day and ironed every night and felt just like the miller's daughter in Rumpelstiltskin, for there was always more. After all, I had been heaping dirty clothes in the extra bedroom ever since September and only washing what we had to have and what I could dry over Stove. After the baby came, I had to wash for her every day, so I threw everything into the extra bedroom. Finally one morning I found the room empty. I tottered back to the kitchen and emptied the wash boiler into the sink and collapsed by Stove. Immediately the sun was obscured by a heavy dark cloud, a wind came swooshing out of the burn; there was a light patter of rain and I fell asleep.

Like coming to the surface after a deep, deep dive, I came at last to the top of my sleep and heard hammering at the back door. I drifted through a heavy mist to the entryway and opened the door. It was the Rawleigh Man, who burst in and snapped me to attention by looking deep into my eyes, and saying, "I heard you got a new baby, organs all back in place O.K.?"

The Rawleigh Man sold spices, hand lotions, patent medicines, coffee, soap, lice powder, flea powder, perfume, chocolate—all kinds of dandy things—and in addition he fancied himself a self-made physician and asked the most intimate and personal questions as he opened his truck and brought out his wares. After I had put his mind at rest about my organs he told me all about his hernia and I'm sure would have showed it to me if I had been a customer of a little longer standing. He told me about a bad ovarian tumor up north, a tipped uterus near "Town," some incurable cases of constipation in the West Valley and a batch of ringworm down near Docktown which had resisted every salve he had.

I made him a cup of coffee and a ham sandwich and he asked me every detail of Anne's birth. He was pleased that I had gone to the "Town" hospital instead of going to the city.

He couldn't have been any more pleased than I, for never in my life have I spent such a delightful two weeks.

The "Town" hospital, run by Sisters, was on a high bluff overlooking the Sound. My room had a ceiling about sixteen feet high, inside shutters on the four tall windows, which faced the sound, old-fashioned curly maple furniture, a bathroom with a chain pull toilet, and pale yellow walls. The Sisters had their own cows, chickens, turkeys, and garden. They baked all of their own bread and thought nothing of bringing a breakfast tray with home-canned raspberries that tasted so fresh I could almost see the dew on them, a thick pink slice of home cured ham, scrambled eggs, hot rolls so feathery I wanted to powder my back with them, hot strong coffee and cream I had to gouge out of the pitcher. They further spoiled me for any other hospital by having homemade ice cream and fried chicken on Sundays and by bringing me tea and hot gingerbread or chocolate cake or rock cookies in the middle of the morning. In the evenings the dear little Sisters brought their sewing to my room and we talked and laughed until the Mother Superior shooed them out and turned out my light. The prospect of two weeks in that heavenly place tempted me to stay pregnant all the rest of my life, but in spite of the coziness of our relationship I did not tell this to the Rawleigh Man.

Other door-to-door sellers were the nursery men, who identified our fruit trees for us and sold us English walnut, filbert, chestnut, apricot and peach trees; the shoe salesmen, who carried no samples, only pictures, and when the brown moccasin-toed oxfords I ordered came, I found out why. The shoes were sturdy—thick-soled—heavy stiff leather—strong sewing—(Gammy would have said they were "baked" together)—firm lining—but they were never intended to be worn. They were so full of tongues and lining and sewing that there was no place for the foot. A person with a more fleshy, less bony foot than mine, might have been able to get one on, but I doubt it. I put trees in them and put them in the closet where they gathered dust until Anne began to crawl. She found them one day and from then on they were her favorite toy. She filled them with blocks and dragged them around like little wagons and a sturdier, more lasting plaything has yet to be devised.

One of the outstanding things about these factory-to-you sellers was their friendly, non-commercial attitude. Money was not important at all. All business was transacted on the cuff and if you had the money in the house when the goods came, fine; if you didn't, you could pay next time. It was so easy and pleasant, with everyone staying for supper or lunch, that we naturally bought more than we needed and in many cases more than we could afford. That was one thing about mail order: you had to send the money with the order and it was hard on people like me who were suckers for deals like "A four-pound jar of Clover Cleansing Cream for only $4.98 —pay when delivered" (or when you can).

There were also a Corset Lady and a Housedress Lady. They travelled together and one squeezed me into a corset and the other jammed me into a housedress. The Corset Lady had piercing black eyes and a large bust and stomach apparently encased in steel, for when I brushed against her it was like bumping into our oil drum. She was such a high-pressure saleswoman that almost before she had turned off the ignition of her car I found myself in my bedroom in my "naked strip" being forced into a foundation garment. First she rolled it up like a life preserver, then I stepped through the leg holes, then she slowly and painfully unrolled it up over my thighs, hips and stomach until she reached my top—then she had me bend over and she slipped straps over my arms and then snapped me to a standing position. My legs were squashed so tightly together I couldn't walk a step and I had to hold my chin up in the air for my bust was in the vicinity of my shoulders.

"Look, Ella," the Corset Lady called to the Housedress Lady, "Don't she look grand?"

The Housedress Lady, who looked just like the Corset Lady except that she had piercing blue eyes, said, "That's a world of improvement, dear. A world!"

I inched over to the mirror and looked. At that time I was thin as a needle and, encased in the foundation garment, I resembled nothing so much as a test tube with something bubbling out the top. Even if I had looked "grand," I had to walk and I wanted to lower my head occasionally, so I took off the foundation garment much more quickly and not nearly so carefully as it had been put on me. The Corset Lady was furious and made no effort to conceal it. While the Housedress Lady was showing me her wares, the Corset Lady sat in a kitchen chair, legs wide apart—but stomach in, bust up—and gazed stonily out the window. Some of the housedresses were quite pretty although electric blue and lavender were the predominating colors, and they were very reasonably priced. I ordered four and two pairs of silk stockings which turned out to be outsize, so I gave them to Mrs. Kettle.

There may have been others, but they were the transients, not the regular door-to-door sellers and not important. I believe that this bringing the store to you, instead of your going to the store, is a fine idea and is a strong factor in breeding contentment. After all, if you know that the Rawleigh Man carries only Field Clover and Wild Rose perfumes, you aren't going to go around whining for Chanel #5; and if you know that the Housedress Lady has nothing but electric blue, you're going to darn well learn to like it or wear feed sacks.

Anyway, it takes the sting out of it if you know that all over the mountains and up and down the valleys all of the women are going to be wearing electric blue housedresses and smelling like Field Clover and Wild Rose.

PART FOUR

Summer

Man works from dawn to setting sun
But woman's work is never done.

12

Who Bothers Whom?

WHO said that wild animals won't bother you if you don't bother them? Whoever said it must have lived in an apartment house and just finished reading *Bambi*. The longer I lived in the mountains the more I realized that Gammy had something when she told us, "Animals are *beasts* and a *wild* animal is a *wild beast!*" Which somehow seemed to make their wildness deliberate and malicious.

Our trouble with wild animals began in the summer. Of course, we had had many encounters with bats, weasels, owls, hawks, wood rats and field mice, but I'm talking now about the large wild animals like bears, cougars, wildcats, skunks, deer and coyotes.

Our summers came early—about May. The sun began to stretch and yawn a little after five and was up, fully dressed and ready to begin a day's work, before six. The days were hot and bright and the house was wrapped up like a Christmas package with roses and honeysuckle whose heavy scent flowed through the windows if the wind was right; and if the wind was wrong I comforted myself with the knowledge that manure was what made the roses and honeysuckle so vigorous and prolific.

The work seemed easier in summer. Washings bleached white and dried quickly, the wood was dry and anxious to ignite, the chicken house floor didn't get soggy, and the paths to all of the outbuildings were clean, dry and hard. It was pleasant to go out in the cool quiet of an early summer morning and hoe the feathery carrots and delicate green ribbons of corn; to transplant lettuce and stake beans; to search hopefully for signs of life in my coldframes until the wonderful smell of percolating coffee hunted me out and warned me it was time to get breakfast.

In summer we made our trips to town in the early morning and were home before the heat of the day. One morning we were fed and scrubbed and in the truck by seven, only to find, after coasting to the county road, that the truck had developed a consumptive cough over night and had become so debilitated it couldn't even make the mild hill back to the house, let alone the vicious grades hemming us in on east and west. For an hour Bob tried to persuade it to go toward town; then in desperation attempted to guide it up the driveway to the garage. It would reel ahead a few steps then slide back, limp and gasping. Finally I was dispatched to retrieve one of the Kettle boys from under one of the Kettle cars to see if he could diagnose the trouble.

I put Anne in her carriage, told Bob to keep an eye on her, cantered down to the Kettles' and persuaded Elwin to come up and fix the car. Elwin ungraciously acquiesced and elected to drive up in the most sinister-looking of all of the jalopies, so I took the trail through the logging burn instead.

For a while the path ran beside the Kettles' stream and was well-travelled and shaded by second growth. Then it became a perilous scramble through giant jackstraw piles of slashings and discarded logs. I threaded my way around brambles, through brush, stepping on logs which either swayed alarmingly over dark bottomless-looking pools or else gave way entirely and left me clinging to slender twigs and feeling for footholds over bramble-filled pits. There were still many great virgin trees left, for this logging had been done by a sloppy small outfit, and the woods were dark and cool and quiet. Occasional birds twittered, and chipmunks slithered over logs and then paused to stare at me glassy-eyed, but there was none of the twig-snapping, brush-rustling, chirping activity that had marked my walk along the road. At last the path miraculously reappeared and wound steeply upward along a ravine and through uncut virgin timber. Halfway up this home stretch I was aware of an uncomfortable feeling as though something were following me. I heard brush crackling across the ravine—even saw branches sway; but when I stopped the noise stopped and finally I convinced myself that I was imagining things. Then I leaped to the ground from a fallen log and there was a terrific crashing across the ravine. Certain that I had not made that much noise, I stopped again to listen, and this time the crashing continued and sounded as if, whatever it was, it was heading across the ravine to *me*. I broke into a lope and at last, panting and scared, I reached home and threw myself on Bob who patted me comfortingly and said that he didn't think that anything would follow me.

Before I could think of a suitable rejoinder Elwin appeared, and while he towed the truck into the yard and started taking it entirely apart, even to removing little tiny nuts and bolts, Bob took Sport and the Kettles' Airedale and went into the woods to look for a good cedar tree to cut into fence posts. He came back almost immediately to get his gun, saying that the dogs seemed uneasy and he thought he'd take a look around.

It seemed hours later when Elwin and I heard the shots. There were four or five close together—then silence. Dead silence. I hallooed. No answer. I began to be frightened and asked Elwin, who was sprawled under the truck, to go out and see what was happening, but he merely stuck his head out, shook his mane of hair out of his eyes, grinned his wide foolish grin, and said, "If he don't come back he's probably dead and there's no use of us both getting kilt, ha, ha, ha!" Then he went back to his tinkering under the truck. After another long wait Elwin came in for a drink of water and, after three dippers full, he wiped his mouth on his sleeve and said, "Well, looks like you're a widder-woman, ha, ha, ha!" I'm sure I would have killed him if the truck had been fixed, but it definitely was not, so I had to content myself with withering looks and a scathing silence.

Bob at last came limping in, his shirt in ribbons, a great jagged bloody gash across his chest, and wearing a beaming smile. "Stepped into the rootpit of a fallen tree and a she-bear jumped me. Fired five shots in her general direction and I guess one of 'em stuck because she's deader than a smelt."

"That gash!" I said weakly.

"Oh, that," he said looking offhandedly down at it. "Must have happened when Joe," he affectionately pulled the ear of the Airedale, "yanked the bear off me. Joe grabbed her hind leg just as she jumped and I guess you could say he saved my life." Then he and Elwin climbed into the jalopy and drove cross country to bring in the carcass.

I got out iodine, bandages, sleeping tablets and my self-control, because, though Bob was being brave and careless in front of Elwin, alone with me, he would act as if the bear had laid open both his lungs and his large intestine, and would spend many happy hours looking for the first signs of blood poisoning. It occurred to me then, that no mention had been made of our dog's part in the fray.

Bob and Elwin returned much later with a large black bear which reeked of iodoform (natural she-bear smell, Elwin said) over the hood of the car and a report of two cubs up a tree. I asked about Sport, but Bob said that he hadn't seen him; that he disappeared just as Joe, the Airedale, got the scent. I looked toward the stove and was relieved to see the dejected tip of a dark red tail. "That's all right, boy," I murmured, "I'll slip you a bone later on just to let you know that I feel the same way about bears."

Bob fixed a sumptuous meal for Joe who was so emaciated that we could follow the progress of each bite. I asked Elwin why Joe was so thin and he brilliantly replied, "I dunno—he should be O.K. We been grainin' him!"

Oh, well, Elwin had fixed the truck, and it ran with purpose and vigor. With the five dollars Bob gave him he said that he was going to buy a fog light for his awful car, which at that time had no lights at all.

After Elwin had left, I hesitantly mentioned the fact that I may have been right in thinking something was following me earlier in the day. Bob said, "Didn't sound like a bear. Could have been, of course—the berries are coming on now—but it was more likely squirrels." With which odious remark he collapsed on his bed of pain and I was allowed to dress the wounds and listen to the stories of the attack.

Now, were we bothering that bear? Of course, some people will say that the woods were the bear's natural domain and just by being there Bob was bothering her. But those woods were *our property!*

I wrote a full account of the bear hunt to the family, and Mother replied that she would love to have the bear skin and Gammy wrote, "Do not leave the clearing"—this made me feel just like a pioneer woman in a long calico dress and a sunbonnet—"please be careful of the baby and why don't you move back to town?"

I had about recovered from the bear, when the blackberries began to ripen. These were the low-growing small blackberries—not the Himalayans which were also plentiful and wild, but came later and did not compare in flavor—and we intended to gather enough for jelly, pies and wine. The summer before we had spent hours in the broiling sun in the old logging works down by the Kettles, getting scratched and stung and burned while we filled five-gallon cans with these elusive blackberries. The resultant jelly and wine were well worth the effort, we thought in the winter when the burns and scratches were healed and summer seemed far away; but here it was blackberry time again and it seemed that even the baby wasn't going to keep me from doing my share, for Bob had found a new picking ground—one where we could take the baby.

One evening after supper, armed with Bob's will power and impeded by two five-gallon cans, two lard buckets, the baby buggy, Sport and the puppy, we trekked a half mile or so through the woods in back of the ranch to a clearing where the blackberries were thick and ripe. Some twenty years before, this clearing had housed a small farm, but the farmer, an elderly bachelor, complained of hearing babies crying in the woods and spent most of his time in town pleading with the sheriff to organize searching parties to find "the little ones." Mrs. Kettle had told us about the bachelor when she heard we intended to pick blackberries back there. She said that the "little ones" crying in the woods were cougars which have a plaintive cry not unlike a lost child and that the woods were alive with them. Bob pooh-poohed this story and said that the old man had probably heard coyotes. Cougars or coyotes, the old man went completely insane and was put away, and no one in that country could be persuaded to move on to his farm. His debtors took his livestock, his neighbors stole his furniture and equipment, and the mountains took over the ranch.

The road which had once connected our farm with the old

bachelor's was completely obliterated by fallen trees, vines, salal and huckleberry bushes, until Bob happened on it early in the spring when he was searching for an approach to a large fallen cedar. He cleared the road enough so that he could get the dragsaw in and haul the wood out, and even I could tell it was a road when he pointed it out to me, because it had two ruts and following it was a little less arduous than just lowering my head and charging through the brush. Pushing a baby buggy over its rooty, spongy, brushy surface was a maneuver which delighted the baby and made me seasick.

From the edge of our potato patch to the second growth, which marked the boundaries of the old farm, was dense virgin timber. The trees, some of them eight and ten feet in diameter, went soaring out of sight and it was dank and shadowy down by their feet. Thick feathery green moss covered the ground and coated the fallen timbers and stumps, which also sported great sword ferns, delicate maidenhair ferns, thick white unhealthy looking lichens and red huckleberry bushes beaded with fruit. The earth was springy, the air quiet except for our grunts and gasps as we extricated the buggy from the grasp of a root or lifted it over a sudden marsh. Bob pointed out the upturned toes of the tree where the bear had attacked him, and I suddenly remembered the cubs, which had disappeared during the night after the death of their mother. Bob stated reassuringly that they were probably still in the immediate vicinity and good-sized bears by that time. It was a distinct relief to reach the second growth of the old clearing and to have the familiar bird and insect noises begin again. Even the air had a different feeling, the pale green silky deep-forest air being replaced by the regular sharp evening-mountain variety. Sport and the puppy, who had been following in stately and unaccustomed dignity, began racing through the brush yapping and yipping at each other, and Bob and I forced the carriage through the waist-high grass to the whitened bones of the cabin and barn where the blackberries were.

When we were finally settled and the berries, as large as thimbles, had begun to plink into the lard buckets, twilight was upon us and so were the mosquitoes, which spiralled out of the woods like smoke and made a direct line for the baby buggy. I slapped and fanned while Bob picked, but it was no use. The mosquitoes crawled into the crevices of the wicker buggy, up Bob's trouser legs, down our necks and inside our sleeves. I took the baby, Sport and the puppy and started home.

Going back through the dusky woods I found it difficult to guide the buggy and at the same time to search the underbrush fore and aft with nearsighted eyes, for cougars, coyotes and bears. It was also disquieting to know that my vision was so myopic that a wild beast would have had to lay its lip against mine and snarl before I could recognize it. In my haste to reach the farm I gave Sport and the puppy the impression that we were having a race and they went bounding out of sight leaving me alone in the gloom. Twigs broke with a loud snap to the left and right of me; overhead the deep quiet was broken by the loud sudden rush of wings. It was most irritating since I was intently trying to listen for a pursuing wild animal. Then just as I reached the point where I was going to scream "Shut up, so I can hear something, will you!" I saw our potato field and our lovely cozy farm buildings dead ahead.

I was putting the baby to bed when Bob came in on the run. "Have the dogs been barking? Have you heard anything?" he breathlessly demanded, as he threw open the closet door and began shuffling through his guns. I showed him Sport and the puppy playing tug-o'-war in the front yard with a piece of rope and asked what sound he had in mind? He snorted impatiently and said, "I was kneeling, picking berries under the leaves [where they hide long and black] and I had that prickly feeling you get when something you can't see is watching you. I looked up and the long grass between me and the woods was waving back together, closing the gap made by the passing of some animal. Some one of the cat family, I'll bet, because there wasn't even a rustle. Not a single sound. It gave me the creeps to think that something had passed so close to me I could have reached out and touched it and it hadn't made a whisper of noise." I thought of my recent trip through the woods with the baby, and the little blood I had left turned so cold my veins recoiled from it in horror.

"Do you think it followed you home?" I quavered, as I began a tour of the house, bolting and locking all of the windows and doors.

Bob disdained to answer such a ridiculous question and began whistling for Sport and pocketing shells. When Sport reluctantly slithered through the door, Bob took him by the collar and they left on an extensive inspection of the premises. I huddled by Stove, afraid to go to bed because I found that the only lock on the back door could have been wrenched off by an anæmic sparrow. I dragged the kitchen stool into the corner and, using it as a table, began a letter home:

"Dear Mother: I am enclosing my sterling salad forks. Please turn them in on some stout bear traps and a good skinning knife. Don't you think it is stretching this wifely duty business a bit too taut to ask me to spend the few minutes of the day when I'm not carrying water, baking, canning, gardening, scrubbing, and taking care of the chickens and pigs, in fending off wild beasts? It is beginning to look as if we are hemmed in by bears and cougars, yet we live in the day of the magnetic eye, automatic hot water heaters and television."

As I sealed the letter, Bob returned and reported small piles of feathers along the pullet runs, but no other tracks or signs. He said in his opinion the animal was a cougar, but it might be just a large wildcat. He also said, "Sport is without a doubt the dumbest animal alive and couldn't follow the scent of a roast duck at twenty paces." Sport hurriedly offered his paw, but Bob turned on his heel and slammed his gun into the rack. I wanted to lock the bedroom windows and have Sport, the puppy and the baby sleep with us, but Bob scoffed at such cowardice and insisted on night air and life as usual. The next morning Bob found prints as large as saucers in the dust of the road directly in front of the house. "It is a cougar and a whopper," he joyfully announced. Then with Sport and his gun he left for the valley to assemble a hunting party. "Am I supposed to handle this cougar with kindness or a hatpin?" I called after him resentfully.

"Would that I had it to do over again," I reflected bitterly as I put a chair under the knob on the kitchen door, "and I would choose the indoor type. Preferably a gambler in a green eye-shade, who sat all day and night indoors under electric lights wearing his delicious pallor which proved that he hadn't been out in the daylight or night air for years and years."

Things got worse. Bob returned from the valley and requested that I feed and water the chickens and gather the eggs while he drove up to some logging camp to get Crowbar and Geoduck Swensen, who were the best hunters in the country and the only ones he wished to trust with his precious cougar. I refused to do the chores on the grounds that I was not going to put my foot outside the house while the cougar was loose. It got very warm in the house with all of the doors and windows locked, but better to drown in a pool of my own sweat, I thought, than be torn limb from limb by a wild beast. Bob slammed through the chores and drove off again. He returned about four-thirty with Crowbar, Geoduck and Crowbar's large bear dog. While Bob finished his evening chores, Crowbar and Geoduck lay in the shade of the truck and gulped down a pint of moonshine each. Then Bob joined them in a third pint and, shouldering their guns, they disappeared into the woods directly back of the pullet runs. Almost immediately the dog began to bark and, guided by his excited yips, I was able to follow the progress of the hunt around the ranch. In spite of my furious threats of the morning, I had to open the windows—the screened ones, of course—both for the breeze and to follow the hunt. A beautiful June bug with gold legs, bronze belly and iridescent peacock-blue back crawled along the screen looking like an animated lapel pin. The kitten caught a baby mole and tortured it on the path to the feed room.

The sun was beginning to yawn and edge toward his bed behind the far mountains, the livestock were making soft contented end-of-the-day sounds, the ducks were taking a last dip in the pond at the foot of the orchard and pre-evening coolness was in the air, when the barkings became wild and concentrated in the woods out by the old spring down past the potato patch. Then there were two shots. Then shouting, barking and at last the hunters appeared bearing the cougar in a stretcher made from their hunting coats and guns. The cougar measured eleven feet from head to tail tip and was the largest ever bagged in our community (according to Crowbar and Geoduck). He was an old-timer, quite grizzled about the head, but with the coldest yellow eyes and the largest sharpest teeth I had ever seen.

Bob was delirious with pride and gave me numerous moon-shiny hugs and kisses and an inch by inch description of the circling of our forty acres, the almost treeing and the actual treeing of the cougar in a small alder, the times the gun was aimed, the time he almost took a shot but decided not to, and at last the closing in and the final shot. I refused several hundred drinks of moonshine from Crowbar and Geoduck, who seemed to be padded with pints and who took long gurgling swigs after each sentence, politely wiped off the lip of the bottle on a filthy sleeve and passed it to me. Bob matched them drink for drink, and the cougar increased in size and ferocity. When they at last reach the point where they didn't care who won and the cougar had been a "shwell shport," they left with the kill, the hunting dog and the guns in the back of the truck. I went to bed with a nice juicy story about love in an apartment house and the bedroom windows opened wide to the velvety night air.

Shortly after the cougar episode, Bob left with the truck one morning to help an East Valley farmer with his haying, and while I was still standing in the drive watching the truck over the brow of the last hill, a skunk strolled in the open back door and settled himself by Stove. I hurriedly shut Sport and the puppy, who were fortunately out by the chicken house, in Sport's yard. Then I tried luring the skunk. I put out a little trail of milk, meat, water and cereal. The skunk blinked and snuggled closer to Stove.

He was actually a civet cat, which is just as evil smelling but smaller than a skunk, but he was very determined. He allowed me to go in and out of the pantry which was across the kitchen from Stove, but one step closer than that brought him to his feet looking menacing.

When Bob finally came home tired and very hungry, I was crouching in the driveway trying to warm the baby's bottle over a smouldering pile of faggots and the skunk was resting in the kitchen. Bob walked loudly and masterfully in the front and skunk swaggered out the back, sneering at me over his shoulder as he disappeared into the woods. He evidently came back that night, for we heard Sport barking loudly in the cellar just under our bedroom, and then the room was filled with that most penetrating, most sickening of odors—skunk. We slept with clothespins on our noses, not only that night but for a week. I bathed Sport with strong soap and Clorox but he remained unpleasant-smelling for weeks.

It just goes to show. In every case the wild animal bothered us first and it was merely luck for our side that Bob was nerveless in emergencies and a crack shot.

13

That Infernal Machine, the Pressure Cooker

TOWARD the end of June when the cougar episode had cooled somewhat, Bob and I made several early morning pilgrimages to the abandoned farm and picked five gallons of wild blackberries—and the canning season was on. How I dreaded it! Jelly, jam, preserves, canned raspberries, blackcaps, peas, spinach, beans, beets, carrots, blackberries, loganberries, wild blackberries, wild raspberries, applesauce, tomatoes, peaches, pears, plums, chickens, venison, beef, clams, salmon, rhubarb, cherries, corn, pickles and prunes. By fall the pantry shelves would groan and creak under nature's bounty and the bitter thing was that we wouldn't be able to eat one tenth of it. Canning is a mental quirk just like any form of hoarding. First you plant too much of everything in the garden; then you waste hours and hours in the boiling sun cultivating; then you buy a pressure cooker and can too much of everything so that it won't be wasted.

Frankly I don't like home-canned anything, and I spent all of my spare time reading up on botulism. Bob, on the other hand, was in the thing heart and soul. He stepped into the pantry, which was larger than most kitchens, and exhibited pure joy at the row on row of shining jars. And I couldn't even crack his complacency when I told him that, although the Hickses were at the time using year before last's canned beef, they were busily preparing to can another one hundred and fifty quarts. Women in that country were judged not by their bulging sweaters, but by their bulging pantries. Husbands unashamedly threw open their pantry doors and dared you to have more of anything.

I reminded Bob, as I began hauling out jars, lids, sugar and the pressure cooker, that the blackberries of the summer before tasted like little nodules of worsted and we still had twenty-five quarts. But he was adamant and so "Heigh-ho and away we go"—the summer canning was on.

I crouched beneath the weight of an insupportable burden every time I went out to the garden. Never have I come face to face with such productivity. Pea vines pregnant with bulging pods; bean poles staggering under big beans, middle-sized beans, little beans and more blossoms; carrots with bare shoulders thrust above the ground to show me they were ready; succulent summer squash and zucchini where it seemed only a matter of an hour ago there were blossoms; and I picked a water bucket full of cherries from *one* lower branch of the old-fashioned late cherry tree that shaded the kitchen.

There was more of everything than we could ever use or preserve and no way to absorb the excess. I tried sending vegetables to our families, but the freight rates and ferry fares and time involved (plus the fact that Seattle has superb waterfront vegetable markets) made this seem rather senseless. I sent great baskets of produce to the Kettles, but with Paw on the road every day imploring the farmers to give him anything they couldn't use, even they had too much. I picked peas and took a shopping bag full to Mrs. Kettle, and was embarrassed and annoyed to find two bushel baskets of them sitting on the back porch, covered with swarms of little flies and obviously rotting. There was no market for this excess since the market gardeners supplied the neighboring towns. I became so conscience stricken by the waste that of my own volition I canned seventy-five quarts of string beans and too late noticed that the new farm journal carried a hair-raising account of the deaths from botulism from eating home-canned string beans.

Birdie Hicks took all the blue ribbons at the county fair for canning. She evidently stayed up all night during the summer and early fall to can, for she would come to call on me at seven-fifteen, crisp and combed and tell me—as her sharp eyes noted that I still had the breakfast dishes and the housework to do, the baby to bathe and feed and my floor to scrub before I could get at my canning—that she had just finished canning thirty-six quarts of corn on the cob, twenty-five quarts of tomatoes, eighty-two quarts of string beans and a five-gallon crock of dill pickles. She canned her peaches in perfect halves, stacked in the jars like the pictures in the canning book. They were perfectly beautiful, but tasted like glue. She canned her tomatoes whole and they came out of the jars firm and pretty, but tasted like nothing. Mother had taught me to put a couple of pits and a little brown sugar with my peaches; plenty of clove, onion and finely chopped celery with tomatoes—and anyway, I like the flavor of open-kettle canned fruit and tomatoes.

By the end of the summer the pullets were laying and Bob was culling the flocks. With no encouragement from me, he decided that, as chicken prices were way down, I should can the culled hens. It appeared to my warped mind that Bob went miles and miles out of his way to figure out things for me to put in jars; that he actively resented a single moment of my time which was not spent eye to pressure gauge, ear to steam cock; that he was forever coming staggering into the kitchen under a bushel basket of something for me to can. My first reaction was homicide, then suicide, and at last tearful resignation.

When he brought in the first three culled hens, I acidly remarked that it wasn't only the cooker which operated under pressure. No answer.

Later, because of my remark, he said that I did it on purpose. I didn't, I swear, but I did feel that God had at last taken pity on me—for the pressure cooker blew up. It was the happiest day of my life, though I might have been killed. A bolt was blown clear through the kitchen door, the walls were dotted with bits of wing and giblet, the floor was swimming in gravy, and the thick cast aluminum lid broke in two and hit the ceiling with such force it left two half moon marks above Stove. I was lyrical with joy. I didn't know how it happened and I cared less. I was free! *Free!* F-R-E-E! After supper as I went humming about the house picking pieces of chicken off the picture frames and from the mirror in the bedroom, Bob eyed me speculatively. Then he picked up the Sears, Roebuck catalogue and began looking for a bigger, quicker and sturdier variety of pressure cooker.

14

This Beautiful Country

AUNTY VIDA took another swallow of coffee, rinsed it around in her mouth as if it were antiseptic, and said, "You have solved the problem of living! You have the answer to happiness! There are thousands of people in this bitter old world who only *hope* some day to achieve by dint of hard work and sacrifice what you and Bob have now!" It was nine o'clock in the morning, and Bob and I had been up since four and had not gone to bed the night before until after twelve. Aunty Vida was just having breakfast. It was that part about others hoping some day to achieve by dint of *hard work* and *sacrifice* what Bob and I already had, that got me. And just what were Bob and I supposed to be doing for sixteen or eighteen hours a day? Weaving flowers in our hair and dancing around a maypole? I was overtired or I wouldn't have paid any attention to Aunty Vida, who was a terrific bore but had a saving grace in that she was very appreciative of "this beautiful country." She was so appreciative of it in fact that when she visited us I used to send Bob out to scoop her up occasionally and bring her in the house to recuperate, since I was afraid that in the white heat of her appreciation she might melt and run into the ground. She took another deep swallow of coffee and garbled on about our pure minds, serene faces, God looking down on us and peace at last, and it was fortunate that Bob called to me just then to come and get the fryers he had killed.

Of course, Aunty Vida was an idiot anyway, but our other guests weren't. They were Bob's immediate family and my immediate family and other charming intelligent people, but they all had the same idea. "Beautiful scenery, magnificent mountains, heavenly food, you fortunate people!" they said as they waddled away from the table. They neglected to note that while they lay in slothful slumber breathing in great draughts of invigorating, smokeless, fumeless, clean, appetite-producing mountain air, Bob and I were cleaning and picking chickens, cleaning clams, burying clam and crab shells, washing dishes, packing eggs, shelling peas and finally dragging to bed at twelve or half past.

"Oh, I just love to wash at an old-fashioned sink," they said. I don't mind washing at an old-fashioned sink, either, when someone else has got up and made an old-fashioned fire and carried and heated some old-fashioned water, and I know that in a day or so I will go back to town and wash off the country grime in an old-fashioned bathtub.

When my family came out to visit for the first time, they were more interested in meeting the Kettles than in exploring our ranch. I took them to call, but poor Mrs. Kettle was overcome with shyness and made us all sit in the parlor and tried so hard to be "reefined" that she only began two sentences with Key-rist. When one of my sisters admired the decorations hanging from the mantel, she said, "Aw I didn't want all that goddamned cr-er-trash hanging there, but the girls insisted." Then she subsided in an agony of embarrassment. I asked her where she had bought the linoleum because I was anxious to get some for my kitchen. She said, "A feller come by about ten years ago and he had samples as pretty as you please and I picked out the pattrun I wanted but when the

Jeezly stuff come it weren't the right color and when that feller come back the next year I told him where he—where he . . . where he-e-e-e—" and Mrs. Kettle turned crimson and left the sentence dangling like the flypaper that hung from the lamp hook.

When introduced to the guests, Elwin closed his eyes tight and he didn't open them until we were leaving. Paw alone retained his savior faire. He came clumping up onto the back porch exuding barnyard odors and good will, and after a few hearty stamps to loosen any loosely caked mud or manure he came charging into the parlor and shook hands heartily with everyone. "Glad to thee you, glad to thee you," he beamed as he settled himself full length on the shiny leather couch. Mother said to Mrs. Kettle, "Do you mind if I smoke?" "Not at all, not at ALL," boomed Paw. "Thmoke A WHOLE CARTOON if you have a mind to. Anyone want a THIGAR?" and he laughed uproariously as he proffered a much-chewed cigar end.

By the time we had the house ready for guests, Gammy had gone to visit her sisters in Colorado, so we were deprived of her reaction to our ranch. However, after the cougar episode I doubt if we could have persuaded her to visit us, even if she had been violently enthusiastic previously, as she had not.

Bob's dress-designing sister and her artist husband came for a week that second summer and they were delightful guests. I clung to them like the smell of frying in an effort to breathe in some of their aura of bright sophistication.

Geoduck Swensen, the angel, preceded their arrival by a few minutes with a gunny sack of Dungeness crabs, a water bucket of Little Neck clams and a bucket of butter clams.

I made the butter clams into fritters for breakfast, staying up until midnight cutting them out of their shells, removing the black part of the neck and the stomach, grinding them and wondering what to do with the shells which smelled so horrible after just a few hours in the sun. I used Mrs. Hicks' recipe for the fritters, which was just an ordinary fritter recipe except that where the recipe asked for two eggs I used either six or twelve depending on the number of people I intended to feed. I also used twice as many ground clams as batter and threw in at the last a handful of fine chopped parsley. For anyone whose only experience with clam fritters has been the big doughy blobs with three specks of clam per blob, which most restaurants serve, I would suggest getting hold of some clams immediately and making yourself a batch. Served with dawn-picked strawberries, strong coffee and Mrs. Hicks' thick yellow cream (which we learned we could buy as whipping cream), clam fritters were not easily forgotten. In fact, thanks to the natural resources of that country, all of the meals were notable. My guests even liked the moonshine which was Maxwell Ford Jefferson's best, and that was good because Jeff was the best moonshiner in our country, having come from a long line of Kentucky moonshine people. Jeff never drank himself, testing his whiskey by the feel of it. He said that the gallon he gave Bob just before our guests arrived felt good and didn't smell too bad.

With all the good food and smooth whiskey Bob's brother-in-law heard the coyotes howl dismally all night; he felt the heavy forbidding nearness of the mountains; he saw the majesty in the unbroken miles and miles and miles of trees, but he saw also the loneliness of such a vista. He reveled in the crystalline beauty of the summer dawn, but he helped me light the fires; he thought the spring water sweet and satisfying, but he helped me carry it; he examined Bob's egg records and was impressed by our hens' performance, by Bob's excellent management, but he said, "The thing that defeats me about a hen is its unresponsiveness. You can pour your heart's blood into their upbringing and all you can hope for is a squawk. You can stroke cats, pet dogs and ride horses, but the only thing you can do with a hen is eat it." Jerry said also, "I think this is an ideal spot to do penance in, but a hell of place to live."

Bob's sister said, "But, Jerry, this moonlight, the mountains, the quiet and the food! It's like something you dream about."

Jerry said, "Uh-huh, but we'd rather have a peanut butter sandwich in Grand Central Station, wouldn't we, Betty?" He also insisted on seeing all of my sketches and said that my water colors had great strength and I must keep on with my painting. I tried to think of something just this side of human sacrifice to show my appreciation.

Bob had always treated my painting as a sort of recurring illness like malaria, and I glanced at him quickly to see what his reaction to Jerry's opinion would be. Bob wasn't even listening—he was reading about round worms in the *Washington Poultryman*.

We took a wonderful trip while Bob's sister and brother-in-law were there. We drove down to Discovery Bay, then up to Port Angeles, then on to Lakes Crescent and Sutherland. It was a trip to prove that occasionally fulfillment exceeds anticipation. We left the baby with Mrs. Hicks, and arranged with Mr. Hicks to gather the eggs and feed the chickens, pig, calf, etc., and were on our way by nine o'clock. It turned out to be a day to treasure, to bring out once in a while to fondle and remember. To begin with, the Discovery Bay road, instead of leaping and twisting around the mountains trying to scare its customers to death as most mountain roads do, took us firmly by the hands, led us between banks of rhododendrons and rows of great cedars and firs, up a gradual slope and around well-banked curves until we reached the top of a mountain and a stoutly fenced lookout point called the Crow's Nest but large enough for trucks to back and turn.

Here all obstructions had been sheared away, and we got out and stood on the brink with nothing but the yellow highway fence between us and the Bay hundreds and hundreds of feet straight down.

We were up so high and the day was so clear that we could see the Straits of Juan de Fuca and Victoria, B. C. I had a strong feeling that if I had brought my glasses I could have spotted London Bridge and the Arc de Triomphe. This bay was named by Captain Vancouver who sailed into its calm waters in 1792 to repair his ship the *Discovery* and, as a reward, named the bay Port Discovery and the small island, which sits sturdily at the entrance, fending off storms and high seas, Protection Island.

The bay is horseshoe-shaped, peacock-blue and beautifully trimmed with white shores and black forests. At the head we watched a tiny lumber train dump its load of matches and go snuffing up the hill again. Sharp and clear came the whistle punk's signals for a skidder somewhere in the mountains back of us. Directly below we could make out infinitesimal beach houses, and a more perilous location I cannot imagine because one pebble carelessly kicked off the top edge could work up enough fury on the way down to smash in a roof. Occasionally on the face of the bluff a brave tree, with toes dug in, leaned against the wind, her hair blowing out straight toward the sea.

After leaving the lookout point the road thoughtfully put up its trees again to shield us from the scarier aspects and before we knew it, we were coasting across tide flats on a bridge. The road didn't leave the water until we had passed Dungeness, the famous crab catchery, and reached the flats of Sequim, a very rich dairy country. We bowled along between well-kept fences, herds of sleek Guernseys, and spacious barns until we reached the top of a long hill and the outskirts of a town. The first thing we knew there was a large PENNY'S sign and we were in Port Angeles. Port Angeles, quite evidently supported by pulp mills and their bad smells, is located on the Straits of Juan de Fuca facing Vancouver Island. It is a beautiful town with all the streets ending at the water's edge, a long spit extending into the Sound like a reaching arm, homes and gardens perched on hills that sweep up steeply from the business district, then obligingly flatten out into plateaus commanding a view embracing Victoria, B. C., the Olympic Mountains and the passing freighters in the deep blue Sound. We had lunch at the best restaurant, which was a regulation chop house with starched white tablecloths and a high-class clientele. Bob, dressed in slacks and sport jacket, looked so devastatingly unfamiliar in the booth beside me that I could hardly eat.

It was almost dusk when we got back to Discovery Bay, but Bob insisted that we stop at the "mansion," a decaying and deserted old estate sprawled along a bluff overlooking Discovery Bay and facing the Crow's Nest. It seems that years and years ago a lumber king for some strange masculine reason thought this spot would be a fine place to bring his young South American bride; but she (and I don't blame her) stayed two months, said to hell with the good neighbor policy and ran home as fast as her little South American legs would carry her. The lumber king, hurt and bewildered closed up the estate and never came back.

The main house, a Victorian grande dame, was prickly with cupolas, little balconies and chimneys. The sagging porches and paneless windows gave it a wrinkled toothless look. Crouching at the back was a huddled mass of servants' quarters with caretakers' cottages, barns and farm buildings across the driveway on the other side. Buildings, orchards and gardens were strung along the edge of the bluff, but so obliterated by second growth, firs, blackberry vines and salal that only by stumbling on broken bits of fence were we able to guess what had been where.

It was quite dark by the time we started through the main house, and the creaking boards, bits of falling plaster and sudden bats kept us hushed and goosefleshy, until Bob, who had stayed behind to examine an old plow, came stamping in, slamming doors and commenting on things in a loud hearty voice. We stepped into a ballroom dappled with shadows and oozing atmosphere and romance, and Bob began pounding on the walls to locate the studs. "With just a little fixing up you could probably house three or four hundred chickens in here," remarked his sister coldly. Bob laughed good-naturedly. "There are some wonderful timbers in this old house," he said. "If I could get it cheap enough it would pay me to tear it down and haul the timbers up to the ranch for a new chicken house." This created such a furor of protest that he stopped being volubly commercial, but while I gazed at the little raised stage at one end of the ballroom and pictured South American musicians playing hot-blooded South American music for the homesick bride, Bob was, I could tell by his face, mentally putting in roosts and nests below the windows.

Upstairs there were endless hallways and about twenty bedrooms, but only one very slender bathroom, with high-stepping gray marble fixtures. The master bedroom across the front of the house had a balcony leaning yearningly toward the water far below. By the time we had reached the second floor the moonlight was pointing up broken steps, spidery corners and cavernous closets and Sister and I were anxious for hot coffee; but Bob, still bold and hearty, made us step out onto the rickety balcony and peer down at the phosphorescent hull of an old sailing vessel lying at the bottom of the bay.

Clutching the fragile railing and crawling with cobwebs and gooseflesh, I expected any moment to feel a hairy hand on my shoulder and to turn around and find Boris Karloff.

Home seemed very cozy with cold fried chicken, hot coffee and all of the buildings one-storied and melting down into the ground.

As our back door closed on each departing guest it closed also on dinner-table conversation and the incentive to arrange flowers and wear nail polish. When the Sister and Brother-in-law left, I laid my sketches beside my first corsage and, grimly shouldering my new pressure cooker, I dropped back into the old groove.

15

Fancywork Versus the Printed Word

WITH my usual bad management, when I moved to the ranch I took with me a box of old school and children's books instead of my own books. At first in loneliness and desperation I read *The Five Little Peppers,* Alden's *Encyclopedia* and *The Way of All Flesh,* separately, together, and alternately over and over. I also read magazines, the newspapers and any and all catalogues. I couldn't borrow books because my neighbors never read. Reading was a sign of laziness, boastfulness and general degradation.

Mountain farm women did fancywork. They embroidered their dishtowels and then bleached them so that they always looked mended. They embroidered their pillowcases with hard, scratchy knots and flowers. They embroidered every stitch their babies wore, and they embroidered, tatted, crocheted and otherwise disfigured their own underclothing, handkerchiefs, doilies, bureau scarves, bedspreads, sheets and napkins. They called it "embroidrying" and said, "I'm going to embroidry me some pillow slips." They were at it from infancy to the grave, but as I don't like embroidery in any form, I resolved that they could cross-stitch me to the cross and I would not learn. I'm the type of female the pioneers were tickled pink to give to the Indians as a hostage.

I wrote long pleading letters to my family to sort out and send on my books, but we were so far from any main roads and the sending of anything over mailing size involved so many people and so many arrangements with bus companies, ferries, individual bus drivers, farmers on the route, that we finally decided against sending the books. Mother promised they would bring them on the first visit. On the first visit they came so loaded with candy, cigarettes, fruit, magazines and presents for the house that I was ashamed to mention the books which they had obviously forgotten. Subsequent visits proved like the first. We corresponded feverishly about the books between visits, sending lists back and forth and fighting via mail over who owned what books, but they were always forgotten. "Left on the front porch!" "Left in the garage!" "Stacked in the front hall!" they lamented, but I knew better. In the first place, most of them were loaned to people—no one ever remembered to whom (like the first edition of Ambrose Bierce's *Devil's Dictionary* which we were never able to track down); and in the second place, packing books is not a chore that anyone undertakes just for the sheer joy of it, so it was always put off until the actual time of departure, then forgotten.

Each time we went to town I looked in vain for a lending library and intended to locate the public library, and each time we returned to the farm with the chicken feed and groceries but without any books. If Bob hadn't parked the car where he did one wet blowy Saturday that first November, we probably never would have found the Booke Stalle crouching on the main street between the cheese factory and the barber shop. "Look," I shouted to Bob excitedly. "A new industry!" and I pointed to the slightly crooked sign timidly spelling out the name. After we had been in the Booke Stalle we realized that though it was a new enterprise—something of a miracle in Town—its opening was very much like putting another bunch of faded flowers on a grave.

The Booke Stalle's door opened the wrong way, so that I was jammed against the wall, fighting for my breath and knocking things off the shelves, before I was decently inside. Miss Wetter, the owner-manageress, exuded Sloan's Liniment and seemed to be trying to gather herself together. She was very thin, some age over thirty-five and had a broken tear duct in her right eye. She continually lifted up her glasses and wiped the eye, pulled up her skirt at the waist, and pulled down her cardigan. She was very deaf and had adenoids. Her stock, her prices and her spirits were very low.

I looked over the stock, which, judging from the titles, had been left her by a deceased relative. There were several lives of Christ, *Brewster's Millions, The Broad Highway* by Jeffery Farnol, *Zoroaster* by Francis Marion Crawford, *The Sheik,* a few of Elinor Glyn, Zane Grey, Kathleen Norris. There were some little books of poetry with covers of brilliantly colored flowers: *My Book of Poems* with pansies on the covers, *Poems I Love* with forget-me-nots, and *Hand in Hand* with daisies. There were also some children's books, some very old histories and a dictionary or two. The only thing I could say for the Booke Stalle's stock was that in comparison *the telephone directory* seemed like very good reading.

I asked for a detective story. My exact words were, "Do you have any detective stories?" Miss Wetter said, "It's bighty dice work—I beet lots of dice people."

I said louder, "Do you have any mystery stories?" She said, "Ad I'b od by owd." So apparently was I. I yelled "CRIME STORIES! MYSTERIES! DETECTIVES!" She shuffled through the drawer in the front of her desk and at last, locating a little notebook, she smiled brightly and said, "A dollar a bonth for two books at a tibe."

So I fished an old envelope out of my purse and wrote out a list of books I wanted, paid my dollar and bought some new magazines. As I left Miss Wetter took off her glasses for the eighth time, dabbed at the watery eye, and remarked enigmatically, "I'b odly od page sevedty-two!"

I felt like replying, "Kid, you're farther behind than you'll ever know."

Two weeks later I went to town again and sought out Miss Wetter. She had installed a smelly coal-oil heater, but other than that it was Act II—same scene, same costume, same books, didn't hear a word I said, and was studying my list as though I had given it to her a half an hour ago instead of fifteen days before. Again I wrote everything down, but I was not absolutely certain that she wasn't also blind.

We continued this way until well into February. Then I begged Bob to go in to the Booke Stalle with me to see what he could do with Miss Wetter. He balked at first, saying that he didn't see of what use he could be unless I wanted him to turn her upside down and shake the books out of her. But I gave him my pleading setter look and in we went. Bob turned on a full one hundred and fifty watts of charm, did not raise his naturally husky voice a quarter tone, and darned if she didn't understand every word he said. With a minimum of eye dabbing and cardigan jerking, she produced two mystery stories, only one of which I had read.

She also told *him,* ignoring me, that she had recently bought out a very prosperous circulating library and was expecting the books in a day or so. Bob was courtly to the point of almost kissing her hand, I was so elated over the coming books that I was graciously able to ignore her ignoring me, and Miss Wetter glowed until I thought her veins would burst their seams.

From that day forward Bob had wonderful luck with Miss Wetter and came home loaded with books and pamphlets on *Making the Small Farm Pay; Coccidiosis, Its Cause and Cure; How Many Chickens Can One Man Handle?* and so forth for him, and the first thing either of them could lay their hands on for me. The mythical library which she was supposed to have purchased failed to materialize while I had traffic with Miss Wetter; or else, as I suspected at the time she told us of the deal, the new library was the twin sister of her own and the new Lives of Christ and *Types of Manure and How To Know Them* melted into her own stock and became indistinguishable. Included in one offering for me, selected by Miss Wetter and delivered by Bob, were *Opera Made Easy for Tiny Tots* and *Tom Brown at Rugby.* All I can say for Miss Wetter is that if her library was circulating I should hate to see one at a standstill.

Late that second summer Miss Wetter sent me a series of articles which she had clipped from some paper (with malice aforethought) about a woman and her very aimless husband who by their own choice lived out of the reaches of civilization on the Pacific Coast. This woman was a very, very good sport about everything including eating seaweed and not having her husband work. She didn't have lights, water, radio, toilet, bathtub, movies, neighbors or money and she just LOVED it. Miss Wetter sent it to me, I'm sure (and perhaps Bob had a finger in it too), for the purpose of bringing to light my own bad sportsmanship. But it was wasted effort.

The articles affected me the same way as a book which Deargrandmother sent Mary and Cleve and me when we were children. It was a slender book with a dark red cover filled with good thoughts and illustrated with steel engravings. There were "Birds in their little nests agree," "How doth the little busy bee," "Many hands make light work," "We live in deeds not years," "Early to bed and early to rise" and others of that ilk. We didn't care much for the book, for the pictures were ugly and the content was dull. Anyway we liked *Slovenly Peter* with its fascinating pictures of children with their eyes coming out and their legs broken off and all of the characters thoroughly bad. But there was one thought and picture in the little red book of good thoughts which absolutely infuriated us. The verse said, "If at first you don't succeed, try, try, again!" and the picture was of a smug, near-together-eyed girl sitting on a little stool, her buttoned, picky-toed shoes crossed primly, sewing doll clothes. The doll dress on which she was sewing was of the round hole for the head stove-pipe sleeve variety, and scattered on the floor around the nasty little girl were dozens of the shapeless dresses which she had apparently spoiled.

By the time I was halfway through the second article by Miss I-Love-Hardships, she had become the near-together-eyed little good-thoughts girl grown up. I don't mind people making the best of inconveniences; in fact, I admired that quality in this woman and the articles would have been fine if she had let it go at that. But no, she had to become so hysterically happy that she made living out of doors in the winter up here sound like a vacation in Tahiti. She said that they didn't even build a shelter—they just slept on the ground so they could be close to nature and had only the trees and the stars for walls and a roof. That was too much. I threw the articles across the room, for *anyone* in this region knows that from the first of September until the last of June we either have to have a roof and four walls or a coating of duck feathers. And if we lay on the ground and looked heavenward, as she said they did, for more than fifteen minutes at a stretch, we'd drown.

I was so incensed by this misrepresentation that I told Mrs. Kettle about it. I said, "This woman said that they lay on the ground outside the year 'round." Mrs. Kettle evidently missed the point, for she said, "Well, she's gotta nerve writin' about it. Mertie Williams laid up outside with Chet Andrews and her old man caught 'em and there was hell to pay. He woulda shot Chet, but he'd been havin' trouble with Mertie anyways, and so he was glad to have the excuse to get her married."

Mrs. Kettle was working on a patchwork quilt. Seated in the black leather rocker around which the floor had been thickly carpeted with newspapers, she was sewing small octagonal pieces of gingham about an inch in diameter, on a clean bleached feed sack. The center of the design was an octagon sewn to a large octagon of plain color; from each point of the center were sewn pieces, and to each of their points more pieces. The edges were carefully turned under and the piece attached to the large octagon and to the feed sack by tiny stitches. From Mrs. Kettle's ample lap cascaded a large finished section of the quilt. It was very attractive. I said as much and Mrs. Kettle said, "I've made one of these here quilts every year since I was married. Got 'em in the closet in the spare room—I figger it'll be something real nice to leave the kids when I die. You'd oughta take up quiltin' 'stead of readin' them damn fool books all the time. Piecin' a quilt is real quietin' work. Here, leave me show you how."

She reached down at her side and produced a clean folded feed sack and a large blue octagon. She threaded me a needle and started me attaching the large octagon to the feed sack. She reached behind her and pulled the coffee pot to the front of the stove, and then we settled down with our sewing. Mrs. Kettle said, "When I set and sew like this I think about things.

When I was first married I was neat and clean and tried to keep my house and my kids clean, but Paw's a awful lazy old bastard and it was fight, fight, fight all the time to get him to fix the fences, clean the barn, wipe his feet, change his clothes, and finally I give it up. I says to myself, 'I can't make Paw change and be neat, so I'll have to change and be dirty, or it'll be fight, fight, fight all our lives,' and so I got easier and easier and found it don't really matter one way or the other. Sure he tracks in manure and he don't clean the barn and last week the cheese factory sent us a warnin' about dirt in the cream, but he's real good-natured and he's never lifted a hand to one of the kids and anyways I don't see that Birdie Hicks is so much better off with her Christly scrubbin' from dawn to dark." She heaved to her feet and said, "Git some of them rock cookies out of that jar in the corner of the pantry and I'll pour the coffee."

Some time later I left for home laden with quilt pieces and full directions for the entire layout. I finished one square after dinner and, although I punctured both hands to pulpy masses and was almost blind, I darted about the house holding up the square to see the effect of piece-quilt walls, piece-quilt curtains, piece-quilt doilies. Bob refused to evidence any enthusiasm over this wonderful new accomplishment of mine and stolidly read aloud from the *American Poultryman* an unusually dull article on coccidiosis, stopping dead at the end of every line regardless of content. The clock on the shelf above the kitchen sink ticked loudly, the stove shifted the position of its wood occasionally. Sport whined plaintively in his sleep, an owl hooted, the coyotes began their nightly howling and the evening droned on. The next night I firmly placed my piece-quilt square in the bottom drawer with "Just call me Myrtle's" cut-out dresses, sorted over Mrs. Hicks' latest contribution of magazines, and settled down happily with a story about a murder in a night club.

Mrs. Hicks kindly bestowed on me all of her old magazines. She brought them up to me well wrapped in her contempt for the printed word. Her excuse for having the nasty things in the house at all was a desire for "receipts and pattrons," and she nearly always managed to not include the end of a serial. In those women's magazines I read thousands of stories about girls named Ricky, Nicky or Sticky and boys named Brent, Kent and Trent. They all earned enormous salaries in advertising agencies, and the girls, as totally unjustified rewards for being both dull and usually disagreeable, had great big apartments full of heat, light and conveniences. The only thing that kept me going, as I read these stories day after day, was the thought that Ricky, Nicky and Sticky would probably get their just deserts whether the authors knew it or not, because Brent, Kent and Trent, in spite of their prodigious business acumen and witty repartee, were American men and average (the author implied) and therefore they would either bore Ricky, Nicky and Sticky to death talking about a little chicken ranch or they would cash in on some of those hundred thousand dollar accounts and buy a little chicken ranch. "Then let's see how gay, how smooth, how burnished-headed you are, girls." I would sneer as I ripped the magazine in two and then carefully put it back together again with transparent tape, so I could read some more stories and get mad the next night.

Verna Marie Jefferson, the moonshiner's wife, sent me up hundreds of *True Story, True Confession, True Love, Dream World,* etc., which she bought for the pictures since she couldn't read. I read them all and was fascinated and ashamed. Here I was complaining and I didn't know what a problem was. Was Bob a thief? Was Bob a murderer? Did my Mother drink? Did my sisters smoke doped cigarettes? Then what was I complaining about?

To supplement my reading I wrote letters. I wrote long letters to everyone I knew and wondered, while I was doing it, why I, who had nothing to say, was able to fill four and five pages while William Lyon Phelps never wrote more than a few meaty lines. Among the letters I received were monthly ones from Deargrandmother addressed Dear Child Bride, which I found intensely annoying, because it brought to mind pictures of a ten-year-old girl in pigtails and bare feet being dragged to the altar by a great hairy brute. I wrote to Mother and demanded that she make Deargrandmother stop addressing me in this depressing way, but Mother characteristically replied, "Why stop her? She enjoys it and it doesn't hurt you." In fact, even mentioning the child bride business to my family was a mistake, for from that time on all of their letters were addressed—and sometimes on the outside of the envelope—Child Bride.

One time Bob went away on business and left me on the ranch alone over night—at least he thought he had. It was summer and sultry. I did the chores while great black clouds surged angrily around the mountaintops and the sky became dark and swollen. Coming back from my last trip to the chicken house and with only the ducks and the pig left to feed, I was surprised to find Elwin Kettle in the yard in one of his old cars with a top. He said, "Maw says there is going to be a storm and she wants you to stay all night at our house. It's O.K. about the chickens. I'll drive you up first thing in the morning. She said to bring the baby's bottle and come on."

I was very touched by her thoughtfulness, but a little apprehensive about sleeping arrangements. I need not have been. Mrs. Kettle took me upstairs to the "spare" room which was immaculate, had a large brass bed with one of the beautiful finished quilts on it for a counterpane; a very pretty braided oval rug on the floor; clean ruffled curtains at the windows; and a large bureau with an embroidered bureau scarf, an oblong pincushion with an embroidered cover on it, a mother-of-pearl dresser set complete with a hair receiver and picture frame (a very deluxe catalogue item) and a vase of large red crepe paper roses. It was very cozy there with the dark clouds outside the window bumping into each other and grumbling menacingly and the wind whining in the tree tops. Mrs. Kettle showed me the closet full of quilts, and the baby shoes and hair of each of the children, the bureau drawers packed with Christmas presents all in their original boxes and never used and consisting mostly of nightgowns with heavy tatted yokes, towel and washrag sets, guest towels and crocheted doilies. By the time we had finished examining everything the storm had broken and the thunder roared and the lightning flashed and the rain hammered relentlessly on the roof over our heads. Mrs. Kettle had to leave me to get out the leak pans since the roof had begun to leak some ten years before. Paw hadn't gotten around to fixing it and each year the winds tore off more shingles and the leaks increased until it had reached a state where she kept a great stack of cans and pans in the upper hall. At the first drop of rain she distributed them over the upstairs. Anne and I were assigned two empty coffee cans—one at the foot of the bed and one in the closet. As I undressed the baby and got her ready for bed, the pink! pink! of the leaks dripping into the cans played a little tune.

After I had given Anne her bottle and settled her for the night, I joined the Kettles in the kitchen. The chores were done and they were all gathered around the kitchen table, reading the local papers and talking about the dance to which the older boys were going and which was seventy miles away. The only car that was running had no lights, and Elwin intended to drive it over the mountain roads by sense of smell, evidently. Maw protested mildly. She said, "Elwin you've turned over three cars on that road and two of 'em are at the bottom of the gulch. You was just lucky you turned over where you did—you go runnin' off the road up by the old logging works and you won't come limpin' home with only a busted arm." Elwin said, "Ah, I know that road like a book." Maw said, "Well, what page was you on when you run off the road the last three times?" Elwin said, "Well, once I had a blowout, and once the axle broke and the other time I skidded."

Paw said, "Jutht remember, thon, you pay for your own funral."

Elwin said sulkily, "Well, what am I supposed to do, climb out of my coffin and go to work until I git enough for the funral?"

Everyone laughed and Paw said, "It ain't no laughin' matter. How about thome thupper, Maw?"

So Maw and I brought to the table great bowls of plain boiled navy beans, boiled macaroni and fried potatoes. Already on the table were pickles, bread, canned peaches and rock cookies. We all had cups of coffee, which was strong, but not venomous, since a fresh pot had been made just after I arrived.

After supper Maw and I "redd" up the dishes, but we couldn't wash them right away for the older boys had to wash (very lightly) and comb their hair at the sink in preparation for the dance. When they had finally left with admonitions to be home before milking time and to drive slowly (wasted breath), Maw and I washed the dishes while Paw and the three little boys took a bicycle apart in one corner of the kitchen. The area back of the stove and around the woodbox showed no evidences of the brooding it had done earlier in the year, but there was a very heady odor in that vicinity because of the wet barn clothes of Paw, the work clothes of the boys steaming behind the stove and the many pairs of work shoes drying on the oven door. Heady odor or not, the Kettles' kitchen had a warm human feeling in comparison with my own clean lonely kitchen further up the mountain. The Kettles, owing no doubt to their struggle for existence, had developed strong family ties; and they had generously, for this stormy evening, allowed me to become one of them. It was "Us Kettles against the world."

Maw and I sewed on her quilt and occasionally put wood in the stove and sneered at everyone who had more worldly goods than the Kettles. "Seventy-two poles Charlie Johnson had to buy to bring 'lectricity in to their ranch and he had to help set 'em too. But did that satisfy Nettie? No. She had to have a 'lectric stove, a worshing machine, a 'lectric iron, a vacuum cleaner. Jeeeeeeesus Keeeeeeerist, I says to her, you'd think you was a invalid. I've worshed for fifteen kids and done it all on a board and with a hand wringer and I ain't in a hospital yet, I says to her."

"I wath up there yethtiddy," Paw said, as he stirred some vile-smelling tire-mending concoction on the stove, "and Charlie wath butchering and I athk him for the thpare ribth becauthe they kilt two pigth and I knowed that the two of them couldn't eat all them thpare ribth, but that thtingy thkunk thaid, 'The reathon I'M BUTCHERING, MR. KETTLE, is becauthe I need the meat,' and I wath tho mad I forgot the egg math I had borried."

Maw said, "Well, I would have told him to take them spare ribs and stuff 'em."

I said, "What else could they do with all of those spare ribs?"

Maw said, "Oh, I suppose they give them to some of Nettie's relations. She's related to every sonofabitch in the country."

I said, "Well, anyway when we butcher Gertrude and Elmer, our pigs, I'll give you all of the spare ribs you can eat."

Maw said, "You and Bob are real good neighbors, Betty, but honest to Gawd some of these bastards don't seem to know what the word neighbor means. Pull that coffee pot to the front of the stove, Paw."

About nine-thirty we all retired. I had on my outing flannel pajamas and was just about to blow out my candle when Maw called to me to come to their bedroom across the hall. I took my candle and tiptoed out into the hall, which ran the full length of the house and from which opened eight doors, not counting my room or Mrs. Kettle's. I was a little hesitant about going in, but Maw was standing by the window in a voluminous outing flannel gown, beckoning to me. "Put your candle on the dresser and come here," she said in a vibrant whisper. She had the window opened wide and the wet night air carried the sweet smells of wet earth and evening scented stock, which grew in a great clump below Mrs. Kettle's window. She said, "Look there, down by the south gate, the lightning struck the old maple."

I looked and made out the outline of half of the maple lying across the road. The other half hung torn and bleeding against the pale summer evening sky. The rain had stopped while we were at supper and all that remained of the storm were the dying tree and the splat, splat, splat of the dripping eaves. Maw stared morbidly at the fallen tree for a few minutes more; then, carefully closing and locking the window, took her candle from the dresser and put it on a chair beside the bed. From the bed came the rhythmic beat of heavy breathing and occasional gulp, gulp, snort of a broken snore. Paw had evidently fallen asleep as soon as he touched the pillow and he must have been very tired for I noted that he wore a felt hat pulled down well over his ears, and the usual layers of dirty underwear and dirty sweaters. Maw sat heavily down on her side of the bed, causing Paw to spring up and slant alarmingly but not to waken. She said, "Paw always wears a hat in bed. He says his head gets cold." I realized suddenly that I had been staring in rude fascination at Paw whose mustache twittered on the ends with each gulp and snort, while his eyebrows drew together menacingly with each indrawn breath. I hastily picked up my candle and left.

As I eased in beside small Anne and laid my cheek on one of Mrs. Kettle's best pillow slips, I knew that I would awaken with a basket of flowers imprinted on my right cheek. "The French knots hurt the worst," I thought drowsily as I snuggled deeper and wondered if Paw had taken off his hip boots.

16

With Bow and Arrow

THE Pacific Coast Indians whom I saw were as unlike the pictures on the Great Northern Railroad calendars as slugs are unlike dragonflies. True, most of the Indians I knew were breeds, but the few full-bloods I saw certainly did not lift me to any pinnacle of artistic ecstasy. The coast Indian is squat, bowlegged, swarthy, flat-faced, broad-nosed, dirty, diseased, ignorant and tricky. There were few exceptions among the many we knew.

Among the exceptions were the Swensen brothers, Clamface, Crowbar and Geoduck. They were Bob's good friends. I couldn't count them as mine, for they had no use for women and were unable to understand Bob's attitude toward me. Bob was such a fine hunter, such a crack shot, so lean and strong and manly; yet when I, merely his wife, asked him for some wood, instead of sneering, "Aw, shut up, old lady," or letting me have a well-deserved left to the chin, he docilely obliged. They were openly disgusted with Bob much of the time. They knocked their wives down for exercise and would no more have considered performing such unmanly tasks as chopping wood or carrying water than they would have entertained the idea of helping with the washing. They brought Bob venison, hundreds of pounds of it, clams, crabs, oysters, pheasant, quail, salmon and whiskey. They sometimes brought Bob unexplained hindquarters of lamb or veal and that second summer they appeared one evening at dinner with an apple box full of smoked salmon bellies. They stamped into the kitchen and plunked the box down in the middle of the floor; then Geoduck with filthy hands lifted out one of the smoked salmon and carefully cut off a strip for Bob to try.

There were times when I had been irritated by their treating me with less consideration than they did Bob, but this was not one of them. I had read of Indians preferring rotten salmon and, although I was reasonably sure that Clamface, Geoduck and Crowbar were more civilized than that, still the dirt on the hand that was fondling the salmon was of at least a week's vintage and God alone knew who had handled the fish during the catching, cleaning and smoking. I grinned hatefully at Bob, as with distended nostrils and curled lips he put the salmon in his mouth. With the first chew, however, the distaste left his face. Of his own volition he went over and cut himself another strip and then cut one for me and insisted that I eat it right then. If that salmon had originally been rotten, then all I can say is that all of us Indians prefer rotten salmon. It was delicious, but I realized with sinking heart that smoking salmon bellies would be added to my canning duties, and in order to learn I would probably have to spend at least a couple of days in Clamface's or his brothers' wigwams, or wherever they lived.

We met the Swensen brothers about a week or so after we moved to the ranch, and as I watched Bob's friendship with them and with other Indians grow, I realized why it is so much easier for a man to adjust himself to new surroundings and people than for a woman. Men are so much less demanding in friendship. A woman wants her friends to be perfect. She sets a pattern, usually a reasonable facsimile of herself, lays a friend out on this pattern and worries and prods at any little qualities which do not coincide with her own image. She simply won't be bothered with anything less than ninety per cent congruity, and will accept the ninety per cent only if the other ten per cent is shaping up nicely and promises accurate conformity within a short time. Friends with glaring lumps or unsmoothable rough places are cast off like ill-fitting garments, and even if this means that the woman has no friends at all, she seems happier than with some imperfect being for whom she would have to make allowances.

A man has a friend, period. He acquires this particular friend because they both like to hunt ducks. The fact that the friend discourses entirely in four letter words, very seldom washes, chews tobacco and spits at random, is drunk a good deal of the time and hates women, in no way affects the friendship. If the man notices these flaws in the perfection of his friend, he notices them casually as he does his friend's height, the color of his eyes, the width of his shoulders; and the friendship continues at an even temperature for years and years and years.

One summer evening when Bob was at a grange meeting and I felt safe and unafraid because it was still daylight at eight o'clock, Geoduck and a friend of his drove into the yard and shouted rudely for Bob. I was packing eggs and had the egg scale and the boxes stacked around me in convenient but confining order, so I shouted back just as rudely, "Bob's not here. He's gone to the grange meeting." Whereupon to my great surprise Geoduck and his friend opened the back door and came staggering into the kitchen—they were both very drunk. The friend was an unpleasant little character with a flat nose, small, very crossed eyes, greasy overalls clinging uncertainly to his pelvic bones and a low forehead across which ran a jagged scar welted high with proud flesh.

"You alone here, eh?" he asked, leering formidably and

rocking on his heels. I looked fiercely at Geoduck, who heretofore had treated me very indifferently but never with hostility. "Geoduck," I said sternly, "Bob has gone to the grange meeting and won't be home until after ten." I waited, but neither Geoduck nor his friend made a move to go. Geoduck looked insolently around the kitchen. "You got this joint fixed up pretty good," he said. I said, "You had better leave, Geoduck, *now!*"

The friend said, "Mebbe we ain't ready to go, eh, Geoduck?" He looked at me with one eye and at Geoduck with the other. I too looked at Geoduck with what I hoped was a pleading expression, but his braggadocio had collapsed and he was smiling foolishly, his eyes glassy.

"How about fixin' us a little somethin' to eat?" said Friend, lurching toward me and knocking against the egg crate so that the delicate scale fell to the floor. My hands were shaking so that I could barely set it up again. Through my mind ran glaring headlines, "Lonely farm scene of tragedy—farmer's young wife raped and beaten!"

"Geoduck," I said, my voice trembling. "Take this man and get out of here." Geoduck laughed a silly giggle.

The drunken friend lurched toward the door of the bedroom where the baby was asleep. That galvanized me to action. I jumped to my feet, knocking over a stack of empty egg crates and a bowl of eggs, ran over to the closet where Bob kept his guns, opened the door, grabbed the first gun I saw, rammed the barrel in Friend's stomach and croaked, "Leave now or shi'll oot!"

Either "shi'll oot" means something in Indian or Geoduck was afraid of the gun, for he seemed to awaken and said, "Aw, come on, Pearl, let's get outta here." They left then, driving through my perennial bed and the unopened rustic gate, and I returned to my egg packing, which eventually soothed my nerves, but cracked a good many eggs. When Bob returned, at a little after ten, I indignantly recounted the experience. Bob didn't act at all alarmed; in fact, he shouted with glee when I dramatically described Geoduck's evil friend and then told him his name was Pearl.

"Pearl!" he said, wiping his eyes. "What a wonderful name for a desperado."

I said fiercely, "Bob, you tell Geoduck that he has to apologize to me or he can never come up here again."

Bob said, "Oh, Betty, he didn't mean any harm. Probably just drunk enough so that he was willing to forget you're a woman and be friendly." Geoduck was Bob's friend, period.

Just a week later Geoduck and Clamface and Crowbar drove up one morning to invite Bob and me to an Indian picnic. I supposed it was in the nature of a peace offering and such a rare gesture from an Indian to a woman that I had to accept. After they had left, Bob said, "I think the Swensens feel that if you go to an Indian gathering and they all get to know you, they won't bother you any more."

I said tartly, "Do you mean that to know her is to love her? And when those drunken savages find out how refined I am they'll come up here to discuss the arts instead of lurching around with rape in their eyes?"

Bob said, "Nonsense, Indians don't go around raping people!"

"Not while I have my trigger finger, they don't," I answered bravely.

The next Sunday morning was the last day of August. It was still and hot and hazy. A wonderful day for a beach picnic—even with Indians. At about eleven o'clock Geoduck and Clamface came to get us. They were both genially drunk and refused to let me provide any food at all, except milk and vegetables for the baby. With more than a few misgivings on my part, we were hastily stowed in the back of their car and taken hurtling down the mountains to Docktown Bay. Here were gathered about twenty families of Indians and part-Indians. The women were on the beach setting tables made of planks stretched between driftwood logs, boiling crabs of which there were five gunny sacks, steaming clams in a wash boiler and carrying platters of fried chicken, bowls of potato salad, and loaves of new bread from the backs of their cars to the beach.

Indian girls ten years old and younger were playing on the beach and minding the babies; little Indian boys were out in boats fishing and catching crabs; and the rest of the gathering, including all girls over ten, were sprawled around the cars drinking moonshine out of gallon jugs (which they held with one finger and crooked in their arms) or home brew.

The water was placidly retreating under a curtain of mist; the tide flats steamed in the hot sun; the little island was partly obscured by the mist, but from its steep banks echoed the happy cries of small explorers; and over all floated the delicious smells of seaweed, clams and driftwood smoke. There were no buckskin dresses or feather headdresses, and from a distance it could have been anyone's picnic. The women were dressed in housedresses and sweaters (mostly maroon), reddish cotton stockings, run-over shoes and eyeshades. All of them wore eyeshades and whether this denoted a racial eye weakness or a weekend special at the crossroads store, I did not learn. The children wore bathing suits and their small brown bodies could have been those of any sun-baked children. It was in the area around the parked cars where the scene was pure unadulterated Indian, and I was anxious to get away from there and down to the beach. I gathered up my baby, the robe and the didy bag, but before I could escape I had to be introduced by Geoduck to some of his friends.

The first was a young couple—the girl small and thin, about nineteen years old, the boy lank haired and doltish. The girl said, as we shook hands, "I-had-Siamese-twins-joined-at-the-breast-bone-and-they-are-pickled-in-a jar-and-on-exhibition-in-New-York!" I said, "How nice!" She said, "And see YEWgene here," she yanked her husband forward. "Lookee," she snapped his head back and pulled open his mouth. "Lookee, no teeth. Want to see where they went?" Of course I did, so she threw open the door of their car and pointed proudly to deep teeth marks in the dashboard. "YEWgene got stewed and run into a tree." She and Eugene stood back proudly, so we could all look. Reluctantly, but at last, Eugene's wife left the teeth marks and YEWgene, and shepherded small Anne and me to the beach. She introduced me to all of the women at once with "Meet Clamface, Geoduck, and Crowbar's friend, Betty." Then she said, "Christ, honey, put the baby down on the beach with the rest of the kids. Lookee they're all O.K." I looked and saw several babies crawling around in the sand and seaweed while older children raced around and over them, knocking them down and kicking sand in their faces. One of the babies was gnawing on a large dead starfish. I pointed this out to YEWgene's wife, but she merely laughed and said, "Christ, I bet that little bugger'll be sick tonight." I walked down and took the starfish away from the baby and threw it out into the water whereupon the baby gave a disappointed howl and bit me on the ankle. A little girl came running and said, "You damn fool, why'n't you git your own starfish."

With my robe and the baby I retired by a large silvery log. The sand was fine and white and there were lovely little fluted pink shells behind the log where they had been tossed by some high winter tide. I picked them up and Anne threw them away, and we were having a pleasant game until a little old Indian woman with a face like a dried fig came and settled herself beside us. She was fascinated by the baby's red curls and kept touching them with her little mummy's claw and saying something quite unintelligible to me. It sounded like "Yawk, yawk, gugh"—but on the other hand it could have been "Gugh, gugh, yawk." Whatever it was, I replied with "Yes, isn't it?" for I was very proud of my baby's bright hair.

Then the little old Indian woman began talking to me in English. She was very deaf and quite senile, but she explained (about seventy-five times) that she was the last of her tribe of Indians: that these Indians had been very warlike and fought all of the time until, in 1855, after a great war with several other tribes, there were only ninety left. "Now only me," she concluded. "All rest marry white and mix blood." She was a friendly little woman and anxious to talk to me, but she had trouble with English and she was so frail and so very old it was difficult. She seemed troubled by the degeneration of her people. She said, "I was a good girl. Just one man for me. No whiskey. Others"—she included the entire picnic —"all bad. Sores come out. Bad arms," and she pointed to a little girl drooping by her mother on a near-by log. The child had one short arm with a two-fingered stub on the end of it. On this cheery note, lunch was announced.

I was careful to eat only the clams and crabs which I had watched a clean woman with sound limbs lift from boiling water. Bob sat beside me and ate heartily of everything—these were his friends.

The Indian men came down from the cars and sprawled around on logs and their wives brought them food. They were all quite drunk but still jovial. They jeered at Bob's cleaning and cracking a crab for me.

After lunch everyone stretched out in the sun and slept, and no attempt was made to clean up the ankle-deep mess of crab shells, clam shells, chicken bones, paper plates, paper cups and other debris.

About two-thirty there was a more or less general awakening and more eating and drinking. Then the men went back to the cars, and the women put away the uneaten food in hampers and baskets. As the afternoon progressed into evening the men became drunker and more noisy and quarrelsome. There were two or three fights, and a few wives were clouted soundly for attempting to interfere. About six o'clock the beach fires were built up high and the children who had been in the water the entire day were rounded up and brought to the fire to dry out. About seven, a feeling of apprehension seemed to permeate the women's group on the beach and moves were made to round up recalcitrant husbands, sons and daughters.

One woman tried to take a jug from her fifteen-year-old son. "Come on, gimme that dirty stuff," she whined. "You get bad like your old man." She clawed for the jug, but the boy held it just out of her reach. Finally in exasperation he said, "Aw, go on, old lady. Git back to the dishes," and placing his open hand in her face he pushed. She sat down hard in the middle of the road and her husband, who was sitting on the running board of a car, just behind the boy, laughed until the tears ran down his cheeks.

One woman pulled her daughter, about twelve years old, from under a car and out of the arms of quite an old man. She pulled the girl to her feet by the hair and still holding the hank of hair turned her face from side to side and slapped her cheeks hard. The girl was so drunk she merely giggled and swayed back and forth, and finally in exasperation the mother tossed her in the back of a car where the girl cuddled down and went to sleep. The old Indian from whom she had been separated, lay under the car, eyes closed, his jug clutched to his chest.

I tried to locate Bob, Clamface, Crowbar or Geoduck, but I was told that they had gone down the beach to do some target shooting. It was growing dark when they finally returned and we left, our party now including the little old lady who was last of her tribe, the mother of the little starfish eater, someone's husband who had passed out on the floor and two wives whose husbands had driven drunkenly off without them.

Geoduck drove home with one wheel part way up the bank on the wrong side of the road. I complained, but he explained profanely and thickly that it helped him guide the car. Knowing that on the road up to our ranch this bank was replaced by a sheer drop of from ten to five hundred feet, I begged Bob to let me walk. Bob said, "Don't worry, honey, Geoduck's a fine driver, aren't you, boy?"

When we arrived home, at long last, Maxwell Jefferson, the moonshiner who had volunteered to do the chores for us, had fed the chickens and animals, had gathered the eggs, had a fire in the stove and coffee in the pot. He carried the baby into the house for me, then somehow he removed Bob from the car and sent the others home. While I put the baby to bed, I could hear him giving Bob a talk on the unreasonableness of a law which "puts a sober God fearin' man in jail foh makin' whiskey, but neveh does nothin' to the goddamn fools who drink it."

The next day I washed all our clothes in Lysol and resolved never again to enter into any form of Indian social life.

Sharkey, the old Indian who lived at Docktown Bay and who gave me my first geoduck, drove up one day to get Bob to help load a ship—they were shy of longshoremen, the stuff they were loading was perishable, and the company was combing the mountains for help. Bob left me the chores and went off with Sharkey, and when he returned he had another bosom Indian friend. Sharkey was over six feet tall and was built on the same general plan as a bulldozer. He had an enormous head, the largest head I've ever seen, and it was our impression when first we knew him that he was a victim of some insidious gland trouble and that his tremendous torso, like his head, was unhealthily large. That was not the case, Bob learned on that longshoring venture. The first cargo to be loaded was sides of beef. A stout plank had been laid from the wharf to the freighter's deck, and Bob and an equally husky longshoreman shouldered a side of beef between them and, with much grunting and maneuvering, walked over the plank to the ship down to wherever they stored the beef. Coming back for the next load, they were amazed to see Sharkey with a side of beef on each shoulder start across the plank. Just as he reached the middle the plank broke, and he and the beef dropped to the water some twelve feet below. Bob fished him out, but Sharkey was so incensed at the company's stupidity in providing such a weak gangplank that he quit then and there and spent the remainder of the day fishing for flounder from the end of the wharf. The boss pleaded with him to come back since he did the work of two strong men, but Sharkey wouldn't even look up. From that day on he and Bob spent many happy hours trolling for salmon, but Sharkey never again would load a ship for anyone.

The Swensens and Sharkey I didn't mind, but I did not like the other Indians, and when they came to call I filled up Stove's reservoir with water and after they had left I scrubbed the house from top to bottom with Lysol. Birdie Hicks the Second, Bob called me.

I didn't care. Little red brothers or not, I didn't like Indians, and the more I saw of them the more I thought what an excellent thing it was to take that beautiful country away from them. They had come a long way from Hiawatha.

17

All Our Kids Have Fits

MY BABY had sun baths, vegetables, meat and cod-liver oil. My neighbors viewed these practices in the same light as charms and asafetida bags. Even though I showed Mrs. Kettle the printed instructions in the Government bulletins I had sent for, she was convinced that it was more crooked work on the part of "them politicians" and would bring my baby to an early grave. Her babies and her children's babies—all the babies she had ever had anything to do with—had been fed pork gravy, mashed potatoes, pickles, beer—and had "fits." The number of "fits" (actually convulsions) a child survived was the measuring stick of the father's virility, the mother's knowledge of dietetics and the child's superior physique.

The Kettles sat around their fire in the evening and Maw would say, "Let's see, was it Charlie or Bertha that had seventeen fits in one day? God, was that a day!" She'd sigh reminiscently and the children would urge her on. "Go on, Maw, tell us. And don't forget the time Elwin was blue in the face for two hours."

Maw would begin. "Let's see, Elwin was most a year old and I was well along in the family way with Ernest. It was a real hot night, and Elwin had had summer complaint bad all day. He was fretful and I took him up right after supper and was rockin' him in the kitchen when all of a sudden he stiffened out and begun to foam and get black and I seen he was in a fit, so I put a wet rag on his head and pretty soon he come out of it. He was all right for about an hour then he got another one—then another—then finally he got one so bad I got scairt and sent Paw for the doctor, but the car had four flat tires—" ("I'd been meanin' to fix them tireth all week," Paw would interrupt.) "And Elwin turned kinda blue and his eyes rolled back and he looked awful, and I thought he was dyin' and I begin to bawl, but Paw filled the washtub with real hot water and he dumped Elwin in clothes and all and just kep' his head outta the water, and after a long while he begun to come out of it. Jesus, that kid was limp—just like a rag doll. And white! He looked just like a hunk of sowbelly."

I would glance at Elwin to see how he reacted to this unflattering description, but like many younger children in a large family, he was so delighted to be the center of attention that he wouldn't have cared if Maw had said he looked like pig's intestines. Elbows on knees, chin propped in his dirty hands, his large blue eyes gazing at Maw through the shock of hair hanging over his face, with the intensity of a sheep dog waiting for a bone, he would sit in happy expectation.

Maw continued: "Well, I finally got that kid ready for bed and then I looked at the kitchen clock"—all heads would turn toward the kitchen clock on the shelf over the sink—"and Jeeeeeeesus Keeeeeeeerist that kid had been in that fit for two hours."

At this Elwin would straighten up and look proudly at the assemblage. He would stand and turn around several times and his family would agree that Elwin was a fine specimen and to think he had been blue for two hours.

One summer morning I pushed Anne's crib on to the front porch and put Anne without any clothes in it for her sun bath. Mrs. Kettle came through the orchard just then to borrow some eggs because the chickens had taken to laying in the deep forest down by the creek. When she saw the baby she was horrified. "You ain't gonta leave that kid out here, are you?" she asked incredulously.

"Of course I am," I said. "She has a sun bath every morning."

Mrs. Kettle said, "Joe's wife has a baby just two days older'n that one and he'd make two of her. She don't give him no sun baths." She poked disapprovingly at Anne's fat, firm dimpled back. "Yup, Jeanie's baby would make two of that one."

Perhaps a week later, I had an opportunity to compare Jeanie's baby with Anne. Joe Kettle was hired to install a gasoline engine and water pump for Bob. Joe arrived about ten o'clock one summer morning—bringing with him to spend the day his wife, Jeanie, and Georgie, their big white baby. Jeanie was a beautiful girl, about nineteen, with soft brown hair, dancing eyes and dimples; Georgie was eight months old and looked as if he had been molded out of dough. He was certainly large and rolling in fat, but he was logy and fretful and every time he squeaked Jeanie ripped open the front of her dress and nursed him. She fed him six times between ten and five, and when she wasn't nursing him she was tossing him in the air, jiggling, tickling, bouncing and shaking him and giving forth a torrential flow of gossip.

When I gave Anne her vegetables, cod-liver oil and applesauce, Jeanie was horrified. She said, "Jesus, kid, I think you're takin' a awful chanct. Georgie don't get nothin' but my milk, a bite of potato and gravy oncet in a while and sometimes a little candy. Look how big and fat he is."

I asked Jeanie if Georgie had ever had a fit. She said, "No, he ain't had one yet, but I guess he will. All kids have fits."

The day was warm and muggy, but Georgie had on knitted bootees, shirt, diaper, flannel petticoat, white petticoat, dress, and knitted jacket; and when Jeanie took him out of doors she put a blanket over his large white head, for babies were not supposed to be exposed to air in any form. When a baby was taken out even on hot summer days, he was bundled up like an Eskimo. His bedroom, usually shared with several other people, was filled with its original quota of air and sealed tight against any intruding draughts.

Apparently the babies liked the life, though, for they lived to grow up; and certainly their lives were more fascinating than those of modern babies with their regular hours, sterile bottles and hands-off policy. The farm baby went where his mother went, and when she had coffee, he got some; when she had beer, so did the baby and he spent many of his happiest hours in the movies, at the dances, or being passed from lap to lap in some warm gossipy kitchen.

That was a great country for babies. People were always having them in spite of home-aborting. More often than not the babies were dirty, runny-nosed and smelly; sometimes they were not bright—but they were fondled and loved just the same. Even the men, who were frequently brutal to their wives and usually cruel to animals, were not ashamed of loving babies and stopped to admire other people's and took their own, drooling and wet, with them when they went calling.

Anne, with her red curls and fat dimpled rosy cheeks, drew an admiring crowd wherever we took her, and it was only owing to my strong will that she did not develop a weakness for pickles, beer, coffee, and "fits."

18

Timbah!

ON SUMMER DAYS while I was out of doors weeding in the garden, picking fruit, gathering vegetables or hanging out a washing, I could hear the short sharp Toot! or Toot, toot! or Toot, toot, toot! from the logging camp nearest us. These toots were the signals given by the "whistle punk" to direct the operations of the skidder bringing in the logs. It was a cheerful sound and made a pleasant break in the great blanket of silence which hung over the mountains on summer days. Occasionally, though, the whistle would give a long mournful wail which lasted for several minutes and meant that a man had been hurt or killed. This sound crept up my back with icy fingers and made me vow I would never let Bob work in the woods, as did many of the other farmers.

All the Kettle boys worked in the woods and they told me gruesome tales of crushed legs, smashed hands, high riggers falling from the tops of great trees, fallers being killed by falling limbs and logging-truck drivers tipping over their trucks and being crushed by their own loads. The Kettles worked for the small outfits that logged with steam donkey engines and hauled their logs to the mills on trucks. The logging camp whose whistle I could hear was a very large concern; they ran three sides—which meant they had three great skidders, to which ran three railroad spurs—so they could log three mountains at a time. Bob had several very good friends among the loggers. There were Tom and Mike Murphy (both since killed in accidents in the woods) who "ran sides" for this logging company. They were actually superintendents. Both were unmarried, very quiet, terrific drinkers and painfully shy. There was also Cecil Morehead, six feet seven inches tall, considered the best "faller" in the country, also unmarried, very quiet, a terrific drinker and painfully shy. Whenever any of these three got drunk enough they might drive up to see us. Once Tom decided he would like an eggnog and came up to ask me if I would make it. Of course I said that I would, whereupon he went out to his car and returned with a water bucket of eggs, a gallon of cream and a gallon of whiskey.

I said, "Do you want me to make enough for the whole camp, Tom?"

"Oh, no, Betty," he said. "I have kind of a headache and thought an eggnog would taste good. I thought I might as well bring enough stuff for us all to have one."

Mike was the same. Sometimes he would come up and bring steaks for me to cook. They were invariably two inches thick and each large enough for six hungry people. Mike always brought two apiece. The day after one such occasion I took one of the steaks to Mrs. Hicks and two to Mrs. Kettle and had to stand helplessly by and watch each good lady place the beautiful tender steaks in a cold skillet over a slow fire with lots of chopped onions and carrots. I knew without being there that by dinner every speck of juice would have been drawn out and the steaks would be gray and chewy like pieces of a thick wet blanket. Once I suggested to Mrs. Kettle that steak put into a very hot pan and cooked over a hot fire was more tender and kept its juices. She said, "Not for me, lady. I've et steaks cooked that way in restaurants and they was all bloody. We likes our meat cooked through. Clean through!"

One time Tom took Bob and me to a poker game at a company house. We watched for a while; then Tom took out a roll of bills about six inches in diameter, peeled off fifty dollars and said mildly to the banker, "She wants to sit in a hand." I drew to an inside straight, made it and won seventy-two dollars. Everyone groaned when I showed them what I had done and several left in disgust. Bob took my place and lost all but three of my seventy-two dollars.

Late that summer, when there was already beginning to be a tingly feeling of fall in the air, Tom invited us to visit the logging camp and to see his "side" in action. I left Anne with Mrs. Hicks, and Bob and I drove through the mountains to the camp where they were logging. On the way we passed barren ugly hills which had once been beautiful green mountains and saw mile after mile of slashings, ugly, dry as tinder and inexcusable. The small companies were careless and wasteful in their logging, but their attempts at destruction were feeble and unimportant compared to the wholesale devastation this company left in its wake.

I was surprised at the size of the camp. It was like a small town. There were stores, bunkhouses, mess halls, equipment sheds, shower houses and offices on one side of the road. On the other were forty or fifty company houses for married men and their families. All of the buildings were brown with white trimmings and many of the houses had white picket fences around their yards.

Tom was waiting for us and introduced us to the general superintendent, the timekeeper and several other officials. Then we climbed aboard the train and rode up into the mountains. The train was a long string of flat cars which hauled logs from the woods to Docktown Bay. We stood on the steps of the cab while the loggers rode on the cars. The skidder was a very large steam donkey run by oil instead of wood—as were the small donkeys—and mounted on track. The skidder had a spar and there was a spar tree in the woods. By means of steel cables and drums the logs were whisked into the air and loaded on the flat cars. There was a man in the cab of the skidder, who, according to the signals from the whistle punk, "backed up easy," "held everything" and "highballed." There may have been other signals I have since forgotten. I watched the choker men and the hooktender fasten the chokers on a log as the hooktender yelled signals to the whistle punk. "Whoo!" shouted the hooktender. The whistle punk snapped his clacker, which was connected to the skidder by an electric wire, and the whistle went "Toot!" The man in the cab let out a little more cable or backed up or did whatever the whistle directed.

When the chokers had been fastened and everything was ready, the chokermen and the hooktenders scrambled back out of the way, the hooktender yelled "Whoo, whoo, whoo!" The Whistle Punk clacked three times, the skidder answered "Toot, toot, toot!" and the great log was jerked into the air, where it swung and swayed for a few minutes. Then away it "highballed" toward the skidder and the train. It was very exciting to watch, but I was scared to death when Tom insisted that I take the electric signal from the whistle punk and operate it myself. I was so nervous that I signaled "highball" when the hooktender wanted a little slack and the chokers were not set. The men down by the log shouted and Tom grabbed the clacker and signaled "hold everything." I could hear the loggers shouting, "Well, of all the goddamned sniveling little . . ." Then Tom called out, "Watch the language, fellows, there's a lady here." I was very embarrassed and glad to leave before the loggers could scramble up to find out "what in hell was going on." As we walked up the road to the train I could hear the muted but vehement cursing of the men when they found out a "woman" had been monkeying with the whistle.

Working in my garden the next day, I heard the familiar "Toot, toot!" from the logging camp and I thought complacently, "A little too much line, jerk her back a foot or so." Later on I heard the mournful wail of the whistle signaling an accident and my distress was even more acute than before, because now I knew more of the men, had seen where they worked, had been shown some of the dangers. But I didn't know until two weeks later that the call had been for our dear friend Cecil who had been hit on the head by a falling limb. He came to see us when he got out of the hospital, his head still swathed in bandages. "Cracked my head like an egg," he told us cheerfully. "That limb hit me so hard on the head it drove my feet six inches into the ground, they tell me. All I remember is shouting 'Timbah!' then waking up in the hospital with a helluva headache." They patched his head with steel plates and, except for a more or less continuous headache, he was as good as new, but his logging days were over.

It was Cecil's idea that we drive to an inlet to see a log chute. He said casually one evening, "I think it would be fun to pack a picnic lunch and drive down tomorrow and watch them chute the logs into the water. Would you like to go, Betty?" Would I like to go? Hah! If he had suggested that we spend the day in the Crossroads cemetery or take a picnic lunch to the town funeral parlor, I would have given an enthusiastic yes. True, our social life had picked up somewhat by the end of that second summer, but I had as yet no need for a date book, for even I could remember that the day was Tuesday and that three weeks from the next Wednesday was Bob's grange meeting and that my next engagement was a Christmas party at the schoolhouse, approximately four months from Friday night.

I packed a lunch of fried chicken, stuffed eggs, tomatoes from the garden, and homemade bread. We stopped at a farm on the way and bought a gallon of ice-cold buttermilk for 10c and a market basket of sun-ripened peaches for 25c. It promised to be a good picnic no matter where we went. The inlet, a natural canal formed by the bed of an extinct glacier, was seventy miles long and about two miles wide. It extended south from Docktown through dense forests, along banks of huge gray stones with gnarled firs springing from their crevices at artistic intervals, past flat sandy flats with willow-fringed streams and lovely little bridges; beside oyster flats, summer camps, small towns and logging works. The road followed the inlet so closely that we were almost driving on the beach, and when we reached our destination, a place where the sand was fine and white and a small stream emptied into the inlet, we parked the car under a willow tree, stepped across the road and were on our picnic ground directly opposite the log chute. The shore on the other side, about a half a mile from us, was a steep bluff down the face of which extended the log chute. The first log came down while I was arranging the baby. I heard a tremendous dull boom, like a far away explosion, and turned around just in time to see a geyser of water shoot into the air for a hundred feet or more, burst like a rocket, fling crystal streams of water in all directions, and subside so slowly it was like watching a slow-motion picture. As the water cleared, the log bobbed up with a circle of ripples which spread and grew until they were washing the beach on our side with small slaps. It was such a tremendous spectacle that it seemed unbelievable we could sit comfortably on the beach eating our chicken and watch log after log come hurtling down. After lunch we went swimming in the lukewarm salt water of the inlet; then drove home in the later afternoon sunlight.

Coming back through the mountains, serene and cool in their dark green robes, I asked Cecil how long he thought our forests would last. He was very pessimistic. "Look," he said. "See those red flags?" I knew; they were planted every two or three miles. "Those flags mean 'Watch out for trucks' and trucks mean a skid road and every skid road means a logging outfit. The smaller the outfit the worse the waste. Improper logging is like a bum shot trying to shoot a certain man in a large crowd. He might get his man the first shot, but he's more likely to shoot two or three dozen innocent people trying to hit the man." I counted twenty-seven red flags on the way home. Some of them may have been old, some may have belonged to pole cutters, but even ten were too many.

Bob was so enthusiastic about logging, loggers, camp life and logging terms, that he asked Cecil if he would show him how to fell an enormous cedar on the back of our place. So one morning they set out, armed with Cecil's double bitted falling axe with narrow deadly sharp blades, and Cecil's falling saw which was so sharp and delicately set that they handled it like a soap bubble. For a while I heard the ring of axe blows, then pounding, then the even droning of the saw. Bob yelled for me to come out where they were. I didn't want to go at all. If both of us got clunked on the head by falling limbs, who would go for help? Who would care for the baby? Anyway, this job was extra dangerous because the tree had a bad lean. The shouting continued, however, so I girded up my loins and hiked out. I found them both standing on springboards which had been inserted in opposite sides of the trunk about five feet from the ground. These were necessary to avoid cutting through the swollen base of the tree. In the east side of the tree a deep cut had been made with the axe. The saw was almost through. The tree was swaying and groaning horribly. I thought it an excellent idea if they both got off those springboards and came back to the house and let the next storm take down the tree, but they laughed hearty man-laughs at me and continued to saw. Suddenly they took the saw out and began chopping vigorously. Then they both jumped down. Cecil shouted "TIMBAH!" and Bob echoed "*Timbaaaaaaaah!*" and with a sound like the indrawn breath of a giant the tree fell. It fell between two virgin firs and parallel to the road, so that it was easily accessible for sawing and hauling. In fact, it fell to the inch where Cecil had said it would. He was a wizard, but he had his broken skull to prove who was really boss.

PART FIVE

Autumn

I saw old Autumn in the misty morn
Stand shadowless like silence, listening
To silence.

—HOOD

19

And Not a Drop to Drink

THE WELL at the back of the place dried up during the spring; the spring at the foot of the orchard disappeared during the summer; and we carried August's and September's water from a spring in a valley eighteen hundred feet from the house if we cut across the burn, a mile and a half by road. I was really glad when the spring dried up, for it meant that Bob hauled the water in the truck in ten gallon cans and I didn't have to feel guilty if he caught me washing my face more than twice a day. Bob was so parsimonious with the water when he was carrying it, that one would have thought we had pitched camp in a dry coulee instead of being permanently settled in the wettest country in the world outside of the Canadian muskeg. "I have to have more water!" was my perpetual cry. "More water?" Bob would shout. "More water? I just carried up two buckets." "I know you did," I would explain patiently. "Two buckets equal twenty quarts. Twenty quarts equal five gallons and the stove reservoir holds five gallons. In addition to filling the reservoir I made coffee and boiled you two eggs, made cereal for the baby and wet my parched lips twice. The water is gone."

With set mouth Bob would go down through the orchard and dip out two more buckets. These would see me through the first tub of baby's washing. There were still the rinsings, the baby's bath, the luncheon tea, the luncheon dishes, the floor scrubbing, the canning, the dinner and the dinner dishes —not to mention occasional hair washings, baths and face washing. For these I carried the water from the spring myself —it was so much easier than explaining.

I estimated that I carried a minimum of sixteen buckets of water a day—sixteen ten-quart buckets or one hundred and sixty quarts a day for about three hundred and sixty days. Is it surprising that my hands were almost dragging on the ground and my shoulders sagged at the sight of anything wet? That I was tortured by mirages of gushing faucets and flushing toilets? I could not believe it when Bob announced casually one fall morning, "I'm going to start laying the pipe for the water system tomorrow." He had been plotting the course, tiling the spring and ordering equipment for a long time now, but none of it had been definite enough to bring running water out of the mirage department. But pipe laying was different. Each day I could actually watch the water being brought nearer and nearer the house—foot by foot.

Then the six hundred gallon water tank arrived, knocked down and looking disappointingly like a bundle of faggots. Bob spent a day out in the woods locating four poles, straight, clear and approximately eighteeen inches at the butt end to support the platform for the tank. I scanned the bathroom fixture section of the catalogues, and Bob decided that the bathroom would have to go where my rhododendrons *sans* taproots were thriving. Did I care? Not a whit. I jerked them up and put them by the woodshed. We were all out for water.

Fall was a wonderful time in the mountains. The sun got up at six, but languorously, without any of her summer fire, and stayed shrouded with sleep until about nine. She shone warmly and brightly then, but I knew it wasn't summer because, though the earth was still warm and the squashes were still blossoming, when I looked heavenward I saw the tops of the trees swimming filmily in mists and the big burn smoked and smouldered with rising fog until noon.

Fall and school were still closely linked in my mind, and I could almost feel the pinch of new school shoes when I saw the first red leaf, heard the first hoarse shouts of fog horns. I remembered last fall when we had driven along a valley road one morning early and had seen the children scrubbed and clutching their lunch boxes, waiting at each gate for the school bus. I wondered if we would still be on the ranch when Anne started to school. I thought what a long day eight o'clock to four-thirty must be for six-year-old first graders. While I was absorbed in such reverie one morning, Bob shouted that the water tower was finished—except for the water. To the casual outside eye it was just a very sturdy, well-constructed platform on which rested a round wooden water tank. To me it was lovelier than the Taj Mahal.

Bob shouted down from the high platform, "I feel like running up an American flag."

I was so excited that I decided to go down and tell Mrs. Kettle about it. In the baby buggy I put Anne, a bucket of extra eggs and a half a chocolate cake and, with Sport and the puppy racing fore and aft, we started down. It was a delightful walk and our cheeks were rosy and our spirits high as we trundled up the last lap to the Kettles' porch. I was startled out of my intent maneuvering of the buggy wheels around axles, stray fenders, car parts and tools, by a terrified roar from Paw Kettle in the barnyard. I turned just in time to see him streak out of the milkhouse into the barn and to see the water tower, which was on a platform about thirty feet high and supported by four straddled spindly legs, give a great groan and collapse with a splintering crash on the milkhouse roof.

A geyser of water flooded the barnyard and frightened an

old Chester White sow and her pigs so that they went right through the discarded bed spring which was part of the barnyard fence and disappeared into the oat field.

After a time things quieted down and Paw came sidling cautiously out of the barn and Elwin called from under his car, "Hey, Paw, you dropped something! Haw, haw, haw!" Maw shuffled down from the back porch and for a while they stood and looked. Then Paw whispered, "The bugger almotht got me! It almotht got me!" Maw said, "For Krissake, what happened?" Mr. Kettle looked belligerently at the hole in the milkhouse roof and at the shattered tank. "All I wanted wath a little piethe of two by four. How would I know the bugger'd collapthe." Mrs. Kettle said, "Paw, what *was* you doing?" Mr. Kettle said, "I needed a little piethe of two by four for the apple bin and I thought the other leg could hold her all right. I only took a piethe about a foot long from that leg by the milkhouthe." Maw said, "Well, I'll be goddamned. It was only a little piece you took out of the water tank support? What in hell did you think would hold it up—air?" She started back toward the house. I went with her. We left Paw still muttering, "It wath only about a foot long. Only a little piethe." Elwin was crowing delightedly, "I knowed what would happen when I seen the old fool sawin'. Haw, haw, haw!"

The next morning before seven Mr. Kettle was at the back door. "I heard you wath inthtalling a water thystem," he said as he scrambled off his wagon and adroitly intercepted Bob's intended escape through the orchard. "And I wondered if perhapth you had a few hundred feet of old pipe you wathn't going to uthe, thome extra fittingth OR THOME LUMBER [preferably four thirty-foot four-by-fours to support a new water tower] AND THEN I WONDERED if you could thpare me a few dayth with the haying. We're awful late thith year, but the BOYTH WON'T HELP AND MAW AND I CAN'T do it all alone." Bob said rather sharply, "Of course I don't have any extra pipe. It was difficult enough to scrape together the money for the eighteen hundred feet we have to have, and how can I help you with the haying when I have almost a quarter of a mile of pipe to bury?"

Paw, not at all nonplussed, thought this over for a moment, then said, "Well, I tell you, Bob, the cream check wath pretty thmall thith month and I jutht thought that perhapth you had thome old pipe or perhapth you ordered too much and I wouldn't want to thee it wathted when I got good uthe for it."

Bob walked away in disgust, but Mr. Kettle didn't seem to mind and waved cheerfully to me as he drove out of the yard. I knew that he would be up the next day for something else.

The pipe burying progressed so slowly that Bob finally had to hire help. Jeff, the moonshiner, sent up a good customer of his who was temporarily out of work. Good Customer was a fine worker and an appreciative eater, but he was very fat and each day after lunch he settled himself in the kitchen rocker, spread a newspaper over his lap, unbuttoned his trousers and fell into a heavy sleep. Of course he was entitled to his lunch hour. He had a right to be comfortable and he tried to be modest; but I felt that that open fly was a slap at my dignity. I spoke to Bob, but he thought it very amusing and said that we were lucky to get Good Customer, buttoned or unbuttoned. I felt the same way before long, for Bob became ill. It was our first bout with illness of any kind outside of bear clawings, smashed toes and other ordinary mishaps. And it was sudden.

One morning when the alarm went off Bob said thickly, "I candt ged up—I'b sick." And there it was—Bob was ill. With anyone else it would have been the common cold; with Bob it was a little-known, very serious illness for which he chose to direct treatment. His pillows were in wads just behind his neck so that his chin was on his chest and his cough sounded much worse than it was. He wouldn't read, preferring to snuffle and stare moodily out the window. He made me take his temperature, which soared to 99°, hourly, and howled with pain when I forced nosedrops up his quivering nostrils. His throat was very sore, he said, and it should have been bleeding from calling to me.

The second night he was in bed, Jeff brought him a gallon of whiskey, and Clamface, Geoduck and Crowbar a quart each. "Whiskey," they told me as they poured themselves large slugs, "will cure anything."

"If," I thought, "it will cure a strong leaning toward homicide, I will drink a pint, neat."

Bob was in bed a week, and Good Customer was so kind, so helpful, during that week, that by the second day, if he had elected to go around stark naked, I wouldn't have cared at all. He chopped me so much fine dry kindling and stacked it in the entryway so conveniently that I had Stove hot, really hot, from four-thirty until ten at night. He not only drove the truck to the valley and brought me up as much water as I could use, but he filled and emptied the wash tubs and carried the clothes basket out to the clothes line for me. He fed the ducks, the pigs, and the turkeys, which we had recently acquired and were fattening; and then he built a sandbox for the baby and drove clear to Docktown Bay for fine white sand to fill it.

When Good Customer first came, I used to sit at the table, my stomach rigid with disgust, as I watched him shovel in his food and knew that he would soon be sprawled in the rocking chair, unbuttoned and unlovely. During that week when Bob was ill, I used to sit at the lunch table soggy with sentiment and watch him shovel in his food and wonder why such a divine creature had never married. I asked him at last. He said, "Lady, I never married because I don't like women. Women drive me crazy. They've got no organization and they go puttering around and never get nothing done. Deliver me from havin' one around all the time." I was very hurt, and it didn't help to hear Bob's hoarse laughter come booming out from the bedroom where he had been listening.

At last the pipe was all buried, the engine was started and one bright fall morning I heard the musical splash and gulp of water being pumped into the tank. The tank was then scrubbed and drained and at last filled, and I stood underneath it and said a little prayer, then went into the kitchen to find that the faucet in the sink had been turned on and water was an inch deep all over the floor. I didn't care—it was in the house.

The next day I walked down to tell Mrs. Kettle about the water and to find out how she had been making out since the collapse of the water tower. What they were doing for water was evident long before I reached the house, for, sitting on an ordinary kitchen stepladder at the site of the old tower, was a fifty-gallon wooden barrel into which the ram was busily pumping about fifty thousand gallons of water. The barnyard was awash, and a white Pekin duck and her goslings were paddling in and out of the tool house. The old sow, which had disappeared into the oat field the day of the crash, had made a lovely wallow just outside the milkhouse door and little waves lapped against the old wagons and discarded furniture as she shifted her weight from side to side. Mrs. Kettle was futilely sweeping mud out of the milkhouse door, but every time the old sow or one of her children moved a fresh wave washed in. I called to her from the edge of the flood and she stopped sweeping long enough to call out, "There's some pitch in the woodbox. Put the coffee pot on and I'll be right up." When she finally came in, flushed and discouraged, she said, "You know that mess down there is only one of the thousands I've been in ever since I met Paw. He's good and all that, but he ain't got system." Which was where she was wrong, of course, for Paw had the perfect system for getting out of any and all work. I hadn't the heart to mention our running water.

It was late October. I awoke one night to the great whooshing of the wind through the forests which meant a storm was gathering. I felt the house give a convulsive shudder as the wind slammed its shutters and rattled the chimneys. The clock ticked loudly—tick, tick, tick, tick. I thought, "This is a winter storm that's coming. Soon it will be winter—a long dreary, gray, wet, lonely winter." The wind gave a derisive howl and dove headlong into the burn to worry the frail old snags. A few drops of rain fell. I felt a tremendous depression settle over me like a sodden comforter. Then from the kitchen I heard a small noise. It was a gentle little sound, but it had penetration and rhythm and soon I could hear it above the storm—above the clock—above the nervous rattling of the house. It was the friendly split, splat, split of a dripping faucet. Our kitchen faucet. That was it—I had water. I had almost forgotten. The winter prospects brightened. Soon I was asleep, lulled and quieted by that supposedly nerve-wracking sound, a dripping faucet.

20

The Root Cellar

A ROOT CELLAR was originally a storage place for root vegetables such as carrots, potatoes, rutabagas, beets and turnips. It used actually to be an earthen pit where the vegetables were buried against the winter freeze. Root cellars in the mountains were more elaborate affairs—built to store fruit and vegetables in winter—milk and cream and butter in summer. Our first root cellar was a rather poorly constructed house next to the feed room. It had shelves and bins of sorts and a dirt floor. During the second spring and summer Bob built a new one, constructed like a mine shaft, tunneled into the bank near the driveway, timbered on top and on the sides, lined with double walls filled with sawdust, with a door that would have done credit to a bank vault, and a floor of white sand. Bob built shelves for all my canned fruit, screening shelves for the winter pears, bins for apples and potatoes, racks for squashes, cabbages, pumpkins and my wrapped green tomatoes, and bins for clean sand for celery and carrots. There were also spaces for crocks of pickles and cupboards for storing milk, butter, cheeses and lard. That storing, storing, storing against the winter should have given me a feeling of warmth and security. It didn't. I felt much more as if I were being prepared for the tomb. I spent so much of that first winter in gooseflesh that it took me the next spring and summer to get my skin to lie down flat again; preparing to go through the whole thing again, no matter how much we had to eat, left me cold. Cold and lonely.

The root cellar began where the pressure cooker left off. I canned my last quart of corn, emptied the pressure cooker, dried it, put it away in the pantry and it was time to put on jeans and help with the potato digging. I love to dig potatoes. To me it is a very exciting occupation, especially when the soil is ideal and each hill yields large, middle-sized and small ones. Our potato crop that second fall was so terrific that we kept track of the potatoes per hill and called excitedly to each other as we broke the record with new hills. The potatoes ran from six to twelve inches in length and two to six inches in breadth—they were free from scab and cooked to a dry white fluff. We had five tons, of which the most perfect were laid aside for seed—the largest put in a bin for baking and about three tons of mediums were sacked and sold.

A month after the potatoes, we dug the carrots, beets, rutabagas, mangels, celery root and parsnips. In between the potatoes and these root vegetables came the apples, pears, squashes, pumpkins and green tomatoes. We left the cabbage, Swiss chard, broccoli and kale in the garden along with the winter spinach, then several inches high, and a fall planting of early peas, but not the celery. We tried it one year and it was pithy and bitter, so along in October we lifted out the celery, being careful to take along lots of damp earth, and buried it in layers in a special dark, damp corner. It kept beautifully.

In September we bought a cider press and I gathered buckets and buckets of windfall apples and set them outside the feed room, where Bob squashed them into cider for vinegar and cider to drink. I made one five-gallon crock of cherry leaf sweet pickles and one five-gallon crock of garlic dill pickles. I found a wild plum tree down at the edge of the big burn loaded with hard green plums which Mrs. Hicks said made wonderful olives. I intended to experiment with perhaps a pint or two but Bob got wind of it and came staggering in with a washtub of plums. They were the size of jumbo green olives and the consistency of bullets. I made a five-gallon crock full and gave the rest to Mrs. Hicks. Five gallons of this, fifty quarts of that—two hundred jars of everything. "At this rate we will have to hire a full-time worker just to unscrew jar lids," I muttered to myself.

But then one clear warm day I walked down into the valley and picked a market basket of field mushrooms and experimented, *of my own volition,* with canning them. My canning book told me how to can mushrooms but said, *"Never can the wild or field mushrooms.* Only an *expert* can tell the difference between a deadly poisonous variety and the common field mushroom." They didn't scare me though. "What is a little mushroom poisoning compared to the dangers of botulism?" I said. "And if we are going to allow ourselves to be scared by a few mushrooms, who is going to eat those thousands and thousands of jars of potential death disguised as string beans, peas, asparagus, beets, carrots, spinach and meat." Ha! I scoffed as I compared a small button mushroom with the picture of the Destroying Angel toadstool which I had propped up beside the pressure cooker. Maybe it was and maybe it was not so I tossed it into the pot.

It was also my idea to gather a water bucket of hazel nuts and a bushel of Oregon grapes. The hazel nuts were wild filberts which grew plentifully along the edge of the roads —the Oregon grapes made a wonderful jelly to serve with game.

Then it was time for the butchering. Earlier in the fall I had begged Bob to take over the feeding of the pigs because I was becoming too fond of Gertrude and Elmer and did not wish to go through the winter bursting into tears every time

I fried the breakfast ham or bacon. Bob thought this was pretty silly of me at first, then he took to telling me of little incidents where the pigs showed great intelligence or affection and the next I knew, I was again feeding the pigs. The day of the butchering I took the baby and walked down to Mrs. Kettle's. That they too had been butchering I could tell from a heap of entrails in the driveway opposite the kitchen door. The entrails hadn't really begun to smell too bad—just very entrailly—but the flies were like a black undulating cloth thrown over the gray shiny pile, and the back porch was heaped with cut meat which even my cursory glance revealed to be dotted with little clumps of fly eggs. It was not appetizing but didn't bother the Kettles as I could tell from the smell of roast pork coming from the kitchen. Mrs. Kettle had just made a fresh pot of coffee and baked a coffee cake and in spite of the entrails, the butchering and the fly eggs I was soon sitting at the kitchen table, eating and drinking with good appetite. The entrails were in my direct line of vision as I sat at the table but still I had another piece of the crusty coffee cake and another cup of coffee while Mrs. Kettle discussed the relentless piling up of the manure around the barn and the ever-present money shortage. "A year ago my stomach would have rebelled actively at first sight of the entrails," I reflected dreamily. I had come a long way in a year—but was I climbing upward toward some sort of well balanced maturity or sliding downhill into a slothful indifference? I asked Bob about it at dinner but he merely looked at me quizzically and went to bed without answering.

The next day we received a long and heartening letter from Bob's sister and husband urging us to buy a place nearer Seattle—what about one of the islands in the Sound, they asked. Bob read the letter and with no comment tossed it back to me. This was Bob's ranch and his work and he was happy in it and he resented any outside interference. I said nothing to Bob but after dinner I wrote a long account of the Indian picnic, moonshiners and the Kettles to both my family and Bob's. Just the actual setting down of the happenings made them lose their portent to me but I hoped they would arouse a little consternation on the other side.

By noon the next day the hams and bacons were soaking in brine; Bob was cutting applewood for the smokehouse; I was trying out lard and grinding up meat for sausage and thinking, "It's all in the mind, anyway, and by adopting the right attitude even I can be cold-blooded and think 'This is just pork—it bears no resemblance to my pigs.'" Then Bob came in bearing Gertrude's and Elmer's heads and asked me to cook them and make head cheese. It was just as if he had brought in Sport's and the puppy's heads and asked me to cook them. I gave an agonized howl and pushed him out of the kitchen and he disgustedly boiled the heads himself in a kettle outside.

I sent a pork roast to Mrs. Kettle and one to Birdie Hicks but there was still more pork than I had ever seen, leering at me from the pantry. Mrs. Hicks volunteered to help me make salt pork for beans and said that everything else which we were not able to eat immediately should be put in the sausage. I thought we should have butchered the pigs one at a time but Bob wanted to get the smoking done all at once.

Under Birdie's supervision, I made the well-seasoned sausage into little cakes and packed them in a great stone crock—first a layer of lard, then a layer of cakes, then a layer of lard. She said that they would keep indefinitely that way; I didn't understand why—still don't—but she was right.

The mountains were getting ready for winter, too. They were very sly about it and tried to look summery and casual but I could tell by their contours that they had slipped on an extra layer of snow—that the misty scarf blowing about that one's head would soon be lying whitely around her neck. With the bright fall weather, the moonlight and the activity on the ranch, I had not yet cringed before their overbearing hauteur but it was there—I could feel it but I didn't care yet.

21

Game or Who Is?

I WAS ALL RIGHT at flushing game, Bob decided, but at retrieving I was a washout. This was due to my near-sightedness and not to any lack of cooperation on my part, I might add. So Bob bought a dog and I uncovered another weak spot in my character.

"This dog," Bob dramatically informed me, as he gingerly untied a large, curly haired, mahogany-colored dog which he had roped in the back of the truck, "is a thoroughbred, Chesapeake retriever, has a pedigree, is a wonderful hunting dog and is very, very vicious." Then, with what I considered an overdose of caution he secured the dog to the feed room door with a hawser large enough to anchor a man-o'-war. During all of this the vicious dog regarded us stonily with pale green eyes and didn't twitch a muscle.

Bob added, as he made a large safe detour around Dog to get his feed pails, "He has bitten almost everyone in Town, but I understand dogs and if I take all the care of him, keep him tied up and just use him for hunting, I think it will be safe enough." The dog trainer went importantly off to the chicken house and Dog and I looked at each other. He had on a handsome studded collar with a name tag but from a safe distance I couldn't make out the name and when I received only the cold pale green stare for my friendly overtures of

"Here Boy" and "Old Fellah," I started on my evening chores, leaving Dog to brood.

During dinner that night Bob told me how, when I was in the hospital, he and the doctor had been discussing hunting and the doctor had told him about Dog. The doctor didn't finally make up his mind to part with the dog until it had bitten two postmen and three delivery boys. Bob talked about building a dogyard with eight-foot wire, the steady nerves it takes to train dogs, the heinous crime of treating dogs like pampered humans—here, I surreptitiously placed my napkin over the puppy who was lounging in my lap—and other firm manly things to do with discipline and hunting, and then he left for the Hickses to see about borrowing the team for disking.

Later, remembering I had not fed my goslings and forgetting about Dog, I heedlessly rushed into the feed room and was scooping up chick feed when I felt a nudge at the back of my knees. I turned and there was Dog the vicious, Dog the terrible, offering me his large feathery paw. We solemnly shook hands and I learned from his nameplate that his name was Sport and from him that he never wagged his tail, was dignified and really very shy. Throwing caution to the winds I untied the rope and took him with me to feed the ducks. There I decided to test out the hunting theory to see if it was as ridiculous as the danger theory. I threw Sport a stick to retrieve and he lalloped after it, then tore down to the edge of the orchard and buried it under the plum tree. I confess that I hugged him for this because now there were two of us who seemed to have had no vocational guidance.

When I heard Bob's car in the drive, I removed Sport from behind the Stove and tied him in the feed room and that was the way things were. Sport knew that I knew that he was neither vicious nor a sport but we decided to let Bob dream on for a while.

When Bob was attacked by the she-bear, Sport just happened to be behind the stove—a little out of breath but trying to look as if he'd been there for hours.

When Bob and Crowbar and Geoduck Swensen and Crowbar's large bear dog were on the scent of the cougar, Sport came out of the woods like a streak of flame and plastered himself to me to shiver and whine. I attributed this to a ferocious attempt at his life on the part of the cougar. Bob said, "More likely a face to face encounter with a squirrel." Bob knew about Sport, then.

Bob, Sport and I went hunting in the fall. Great sport for Bob and Sport—all work and no credit for me. The procedure was for us to start up the road with Sport disappearing into the woods at intervals supposedly to flush a covey of quail or some grouse. He would crash around like a bulldozer for a while and then appear all smiles and minus anything at which to shoot, except him, which proved a stronger and stronger temptation as the day wore on. Finally I would cut through the brush and flush some grouse, which left the ground just in front of me with a roar of wings that scared me so I fell over a log into some blackberry vines. Bob would take aim and fire and usually get one or two birds which always fell in a very dense thicket. Wallowing in, bent double, in an attempt to see the birds, with Sport rushing between my legs and back and forth just in front of me, I would inadvertently stumble on the first grouse, the other being only a few feet away. Thoroughly scratched by blackberries, stinging from nettles and small whipping twigs, I would reach for the first grouse but Sport had found it too and, snapping it up, bounded off to find Bob. I could hear Bob praising and patting as I crawled under a log with the salal snagging my cheeks and blackberries wrapping themselves around my legs. Emerging at last I would give Sport a dirty look as Bob calmly and without comment took my grouse and stowed it in his hunting jacket.

Bob really tried to train Sport. Every afternoon for a week or so during the early summer he'd work on him. Through my open kitchen windows I could hear violent disagreement between the dog trainer and the hunter. From the kitchen it sounded as though Sport wanted to sit heavily on his tail and shake hands but Bob wanted him to take cover, or uncover or something quite different. "Jeeeeeeee . . . zuz . . . how could anything be so dumb?" screamed the exasperated trainer as Sport eagerly offered the other paw.

His breed, color, build and pedigree said that Sport was a hunting dog—he wasn't—he was a friendly dog. He loved companionship, warm fires, the baby and me. He preferred chocolates to dog biscuits or meat and he was passionately fond of music. I still bear the emotional scars of the first time I heard Sport howl. It was the first night of the harvest moon and I was lying contentedly in bed watching the silvery moonlight on the pond when suddenly the quiet air was ripped to shreds by the most terrifying noise. It sounded as though a freighter had gone aground on the front porch and was giving a long hoarse shout for help—it sounded like an ape man roaring for his mate—it sounded like the death call of an elephant. It woke Bob from a sound sleep and sent him flying for his gun. Just as he reached the window it came again—louder—more terrifying. Bob thrust his head and shoulders through the open window—then he burst into roars of laughter. "Come here, Betty," he said. I went over to the window and there below us sitting in a shimmering pool of moonlight was Sport, looking as self-conscious and embarrassed as only a Chesapeake who had built up a terrible reputation of fierceness could look when he was caught wallowing in moonlight and howling for love.

I was, still am, a strong swimmer but Bob took neither Sport nor me on his duck-hunting expeditions. For this sport he either went to the Dungeness marshes, where he got teal and canvasbacks which tasted like fish and were perfect stinkers to pick, or he and Geoduck and Clamface went to a secret private preserve they had and shot mallards which were very very good but stinkers to pick. The private preserve was a lake owned by a Mrs. Peterson, and known as Peterson's Lake. Mrs. Peterson loved ducks and spent all of her pension money buying feed for the thousands which lived on her lake. When she first bought the lake there were no ducks but she wanted ducks more than anything, she told Geoduck and Clamface who happened to be hunting up there. Geoduck on the spur of the moment told her that if she would paint her house green like a mallard, the ducks would come there to live, for green was their favorite color. So she painted her house mallard green and the next year, nobody knows why, thousands of mallards took up residence on her lake. She was hysterically grateful and told Geoduck and Clamface that from that day forward they, and their friends, could hunt ducks there. She shot, quite accurately, at anyone else who set foot on her property.

Her mallards were fat and friendly and Bob preferred the salt marshes for duckhunting but I preferred the mallards. Roast mallard duck, wild rice and Oregon grape jelly made a superb combination.

Sometimes, in fact many times, hunters became lost in the mountains. Even the experts like the Swensens got lost on occasion. Our first meeting with Crowbar was the result of his being lost in the woods near us. It was that first November and I had not yet become used to the coyotes howling and this night I had lain awake for hours and hours listening to that morbid sound, when suddenly I heard two shots—close together—from the forest west of us. I waited a while—then it came again—two shots. I waked Bob and told him. He immediately got his gun and fired an answering shot. The two shots came again, Bob answered with one, then he started the fire and put a pot of coffee on. After quite a while we heard the shots again much nearer—Bob answered. This went on through two pots of coffee and all of my morning's wood and I think that Bob and I were more tired than the hunter when about three-thirty Crowbar Swensen, who was supposed to know those mountains better than the deer, came wearily into the kitchen. He said he had spent the fore part of the night in a tree and would have been content to stay there until dawn, if he hadn't discovered that a cougar had bedded down just below him the night before. Crowbar was dripping wet and chattering with the cold so Bob poured him a water glass of the whiskey we had bought in Town and offered him dry clothes. He gulped the whiskey without a tremor or a chaser, scoffed at the coffee and dry clothes and the moment the first vestige of dawn appeared he left to pack in his deer.

It was just a year later when Crowbar took us for a drive to show us the "best deerhunting country in the world." We drove down into the foothills and then took a single-track, very bumpy road through miles and miles and miles of burned-off land. "This here's a real burn," Crowbar told us taking both hands off the steering wheel to encompass the entire bleak landscape in one grand gesture. "This here was the biggest forest fire in the whole world." He could have been right, too, for anything in that country to be large enough to be noticed would have to be the biggest in the world. The moon was up, but a heavy mist had begun to settle, so that it seemed as if we were driving through an endless swamp with the silvery water rising to the knees of the millions and millions of stark snags and skeleton trees which covered the hills and stood quietly on either side of the road. The road dipped into hollows and rose to the crest of hills but we seemed to be getting nowhere, for the scene remained the same. I tried locating a particularly tall snag on a distant hill to use as a marker in our progress but I always lost it. There were so terribly many of those tall snags. So many desolate hills. There was no sign of habitation. No sign of life although Crowbar assured us that in the daytime the hills, even the roads, were alive with deer. On and on we drove past lowlands and hollows filled with the seething gray mist, past the spiked hills outlined by the pale moonlight. The dampness was penetrating and the night grew colder and I was thinking, "This is what Purgatory must be like," when Crowbar at last decided that we had seen enough, turned the car around and we started home. Even bright sunlight, blue skies and bird song would have had a hard time lightening that landscape.

22

The Theatah—The Dahnse!

MOST social gatherings I can do without, I have said. By that I meant all women's teas, luncheons, evening bridge parties, all New Year's Eve celebrations, and all large parties. I lived to eat those words—in fact I lived to eat them over and over again like a cud. For, by the time I had lived on the chicken ranch for a couple of years I would have crawled on my hands and knees over broken glass to attend the Annual Reunion of the Congenital Idiots' Association.

Entertainment offered us chicken ranchers was: Saturday night dances—from twenty-seven to seventy-five miles away; moving pictures in near-by towns, seasonal entertainments at the schools—interesting only to the parents of the participants; monthly basket socials given by the churches—these socials were owned and operated by Birdie Hicks and her ilk and were for the sole purpose of gossiping and eating; occasional private parties such as the Indian picnic or the birthday party for Mrs. Kettle; and the county fairs.

Bob and I referred to anything social as either "the theatah" or "the dahnse" due to an unusual contact we had made during our first year on the ranch. It was a bright fall day and we had been out trolling for salmon. We had just returned Sharkey's boat and were walking down the beach to the car when we noticed a woman and some children digging clams. The woman was clad in a few shreds of hip boots, men's trousers, a gunny-sack apron, a purple knitted shawl, and had two long, dusty brown braids swinging from her head. With her were four children—a couple of them by necessity on all fours. They were all filthy and not bright. It was not an inspiring encounter and Bob and I quickened our pace, but to no avail, for the woman straightened up and hallooed to us. We stopped and she hurried up. "I am Mrs. Weatherby," she said with a great air. "And this is Mary Elizabeth," she indicated the biggest and blackest of the children, "Pamela Lorraine"—this one was a drooling crawler, "John Frederick"—his eyes only opened a tiny crack so he had to tip his head way back to see us, "and baby, Charles Lawrence." Charles Lawrence was eating a raw clam. We started to introduce ourselves but she waived it, with "Oh, my goodness, I've heard all about you. I knew the day you moved out here. I even know," she waved a dirty finger at me, "that you are expecting." (My God, I thought, I hoped it won't be marked.) "Now the children and I were just going up to the house for a bite and we want you to come along with us. We simply won't take no!" she laughed, batting her eyes and exposing her bad teeth. I was fascinated by her and wanted to go; Bob seemed unable to cope with the situation and numbly followed. Mrs. Weatherby tripped along in front of us, occasionally pausing to scoop one of the children out of the brush and back onto the path which led up a bank from the beach and into the second growth. The path was littered with empty cans, bottles, discarded clothing, papers, empty boxes, broken furniture and other junk. The litter grew thicker as we approached the house until we were picking our way through a sizable dump. The path ended, we looked up and there was the house leering drunkenly at us from its unsteady stilt legs, more cans and bottles pouring from its open door like lava from a crater. "I hope you'll excuse us," called Mrs. Weatherby from the doorway, "I just couldn't bear to be cooped up in the house on such a lovely day, so I left my housework and took the children to the beach." They must have been gone for about five years, I thought, judging by the little household tasks which had accumulated in her absence. The tin sink was so covered by dirty dishes and pans and cans and bottles that it was possible to tell it was a sink only by the dripping faucet. The oven door was gone from the rusty range and the floor, drainboards, chairs, table, washbench and an old army cot were hidden entirely by newspapers and magazines. Mrs. Weatherby took a heap of newspapers from two rickety chairs, kicked the cans in front of them under the sink and bade us welcome. "It's been so long since I've entertained anyone but the local folk," she said gaily, "that we're going to have a real treat." She bent over and began fumbling under the sink and finally emerged with a bottle. "It's only Chianti, but it *is* wine," she laughed and I noticed that one of the heavy dusty braids had dipped in something under the sink for it was wet at the end and dripped. She got jelly glasses from a shelf and filled them, giving each of the slobbering children a little of the wine mixed with water. She said, "The little French children have wine so why shouldn't Mary Elizabeth, Pamela Lorraine, John Frederick and baby Charles Lawrence, even though they aren't French?" [or children, I added grimly to myself]. It was a grisly performance because the children were actually drooling idiots, but not so grisly as what was to come. While Bob and I held our wine, trying to gather up enough courage to drink it, Mrs. Weatherby got a stool out of a corner, knocked the carton full of magazines from it to the floor, climbed up on it and began rummaging around in a high cupboard. As she rustled around up there, I stole a look at Bob. He sat staring at the wine and looking like a man who has just been clipped by an iron pipe. "Oh, here it is," squealed Mrs. Weatherby finally, and down she came clutching a big gray bundle in her dirty hands. She began slowly and carefully to unroll it, meanwhile telling us the secret. "I made it myself last year [she said yeah] but there was no occasion to warrant my using it, so I wrapped it up and tucked it away and how glad I am, for fruitcake is so improved with age." She had finished the unrolling and at first I thought—"It can't be!—it can't be!" but it was. The fruitcake had been rolled in a rather soiled suit of someone's long underwear. The cake was small and dark and I prayed that there would not be enough to go around, but Mrs. Weatherby served Bob and me before giving a thin slice to each of the droolers. Then, seating herself on the stool, one elbow on the dirty drainboard of the sink, one hand daintily holding the fruitcake, the fingers of the other carefully encasing the glass of wine, Mrs. Weatherby told us about the "quaint folk" of that country. She said, "When I first moved up here among these folk, I was dreadfully distressed by the ignorance, the complete absence of any form of culture. I said to Mr. Weatherby, 'I don't know how I'm going to go through a winter without the symphony, the theatah or the dahnse.'" She batted her eyes and nibbled at her fruitcake, waiting for my reaction. I said, "What did you do?" She said, "Why, my deah, I organized a study group. We were to meet every other Thursday—light refreshments, you know—and we would study and discuss. Music I thought would be the first subject—just simple little melodies at first—then a little stronger, and a little stronger, until we were at last able to digest a whole symphony. I had rather an ex-tensive course planned but it fizzled out. I learned to my sorrow," she took a sip of wine, "that these folk are truly simple children of the soil and they wish to stay that way." I could imagine where Mrs. Kettle had told Mrs. Weatherby to put her symphony and her study group. Mrs. Weatherby continued, "And there is no spirit of cooperation at all. Time and again I have offered to help them put on little things at the schoolhouse, but they are still angry over the hot lunch ruckus and won't have anything to do with me." She waited for her cue so I said, "What hot lunch ruckus?" She said, "Surely you have heard?" "No," I answered. "Well," said Mrs. Weatherby, "it was at one of the grange meetings and the women were talking about having a hot lunch served at school. I said, and I had every right, that I could not approve such a program until I had inspected the school kitchens to be sure they were sanitary. Honestly," said Mrs. Weatherby, "from the uproar that ensued you would have thought I had questioned the ladies' legitimacy." I knew that if I caught Bob's eye I would explode, so I stood and thanked Mrs. Weatherby for the refreshments and we left.

When we got in the car, Bob said, "That woman! Completely nuts!" But I had an awful foreboding that given time *I* could be like Mrs. Weatherby, without the idiots of course. Where had she come from? Who was Mr. Weatherby? What had happened to him? Mrs. Kettle, when questioned, said, "Nobody knows where she come from. Sharkey says she washed up on a high tide. She's married to the most worthless, drunken Indian in the whole country and he beats her unconscious every Saturday night. She never tells nothing about herself. When she first come over here she put on lots of airs and tried to teach us about music—'the theatah and the dahnse' [Mrs. Kettle waved the stove lifter like a baton and half closed her eyes in great disdain]—but her neighbors wouldn't have none of that stuff and now she just stays home in that dump or takes them kids down to play with the clams." Poor, poor Mrs. Weatherby with nothing in her life but those names, Mary Elizabeth, Pamela Lorraine, John Frederick and Charles Lawrence. How she rolled them around on her tongue, savoring them to the last drop. I tried rolling Anne Elizabeth around on my tongue but it was too plain and easy. I was safe for a little while yet.

My first contact with the "theatah" was the grade school Christmas entertainment which I attended with Mrs. Hicks. The entertainment was notable for a smell of manure which came in with and hovered over the audience like a swarm of insects and a snowflake waltz danced by ten little jet black Indians whose only claim to snowiness was the whites of their eyes.

Our next brush with the entertainment world was "the dahnse." "The dahnse" was a country dance held every Saturday night, week in, week out, year in and year out. Sometimes the dance was twenty-seven miles away; sometimes seventy-five miles over tortuous mountain roads, but everyone attended. The farmers and their wives and families went because they were Grange members and often the different granges sponsored the dances. The farmers and their wives danced, helped with and provided the food, and behaved themselves. The loggers and millworkers went to the dances to get drunk. They wouldn't for a moment have considered dancing and, after standing on the sidelines and shyly watching for a while, they'd go outside, get uproariously drunk, usually manage to get in a fight, and toward morning could be found either asleep in their cars or in an all night restaurant in the nearest town. The Indians danced, drank and fought at the dances. They were also the cause of the ruling that once inside you had to stay or pay again. "This," explained Mrs. Hicks who escorted us to our first dance, "is to keep couples from dancing one dance and then going out to their cars to drink and you-know-what." "No, what?" asked Bob wickedly.

It was late September of the second year before we felt the

need or the urge to go to a dance, and even then it was not Bob's idea, but "A dance every two years or so won't spoil you for the simple life," he laughed and told the Hickses we would go with them. Having never been to a country dance in all of my life I didn't know what to wear but when I asked Bob he got his voile and leghorn look, so I wore a suit and my hurty alligator pumps and realized as soon as we arrived that I could have worn anything from a housedress to a crumpled taffeta party dress. Mrs. Hicks had on a blue flowered print, a touch of orange lipstick, stiff dippy waves low on her forehead and lots of bright pink "rooje" scrubbed into her cheeks. Her stockings were deep orange service weight and she had her feet encased in long black patent leather slippers with high heels which made her walk with her behind stuck out. Mr. Hicks wore his town suit, the tightest collar in the world and a remarkable hair-do wherein the hair was parted in the middle, wet down thoroughly, while from each side a small group of front hairs were laid back as though they were to be fastened with barrettes. It gave him a look of careful elegance which was unfortunately marred by a strong smell of Lysol. Mrs. Hicks probably made him put it in his bathwater. Mrs. Hicks was wearing Rawleigh #5 and Lysol. Bob was well-groomed, divinely handsome and a little drunk.

We arrived at the dance hall, a large square building, at 8:30 exactly. There seemed to be hundreds of cars from which people were disembarking while shouting to people in other cars, calling anxiously to children who were climbing on radiators, over fenders and into other cars. There was apparently no system to the parking—you just drove into the yard wherever you could, and turned off the motor. Mrs. Hicks, however, directed Mr. Hicks to a spot at the back of the hall near a side road so that we could leave whenever we wanted to. We walked around to the front past several parked cars where couples were already in the back seat drinking, and up on the porch where the tickets were sold. As soon as Bob had the tickets he gave them to a doorman who stamped PAID in purple ink on the backs of our hands and we went inside. The dance hall was very large and brillantly lighted; festooned with dusty green and pink crepe paper streamers; heated to approximately 90° and packed with dancers. The orchestra on a raised platform in the center of the hall consisted of a piano, accordion, violin, trumpet and saxophone manned by hard-working sweaty musicians. The music was very loud. Along each of the four walls ran a bench piled high with coats and as I discovered later, sleeping children of various ages. During intermissions I watched mothers shuffle through the coats until they found their child, hurriedly change its diapers and as often as not, whip out a large breast and nurse the baby through the next dance.

We found a vacant place on a bench, took off our coats and folded them, Birdie Hicks scrubbed a great deal more of the "rooje" into her cheeks and we were ready. Bob and I danced the first dance which was one of those double quick hops that are the delight of country dancers. We stood it for a few minutes, then retired to the sidelines. While we watched the twirling and hopping Bob warned me that I was to dance with anyone who asked me as even a faint display of choosiness would brand me as a snob and immediately and irrevocably ruin my social career. As that dance ended Bob wandered over to speak to some of his logger friends and I stood ready to obey his wishes and dance with the first person who asked me. Of course the first person who asked me was a tiny Filipino, Manuel Lizardo, who rolled his r's so that I could barely understand him and was so short I could hardly hear him. He had his trousers pulled up to his armpits and we must have made a striking couple as we danced around the floor, me crouching like a broody hen over Manuel who, with head thrown back and teeth all exposed was entertaining me with witticisms such as "The musssssss-eeeeek issssss ssssso loud it sssssounds like barrrrrrrrrrnyarrrrrd ahn—ee—mahls. Haa-ha-ha-ha-ha! Ho-ho-ho-ho!" After he had stood on his tiptoes and screamed this at me for the fourth time I understood and laughed dutifully. Bob oozed by like an oil slick with a very beautiful, bad-looking Indian girl, plastered so tightly to him that I expected to see the print of her dress rubbed off on his jacket when they disentangled. I tried to catch his eye but he was intent on his work and didn't look up.

As each dance ended the dancers joined hands and marched around and around the floor until the music began again or there was an exchange of partners. By the end of the third straight dance Manuel had lost his smile and given up the wit and I was wondering what every other woman at the dance had which I seemed to lack, when Mr. Hicks came to our rescue. He grabbed me firmly around the waist, leaned forward and strode purposefully around the hall. The result was that I was bending way over backward, doing little running steps on my toes between Mr. Hicks' legs and expecting momentarily to do a back flip over his iron arm. The music stopped at last and I begged to sit out the next dance. Mr. Hicks gallantly offered to sit out with me but was immediately claimed by Birdie's mother who came skipping up in a pink taffeta dress and little dirty white canvas strap slippers. "Can't bear to sit out a single dance—too full of pep!" she tittered, as Mr. Hicks set his vise and strode away. As they passed by the first time I saw that Mother had relaxed like a scarf over the arm, her little fluffy head bobbed spasmodically and the little dirty white canvas slippers made jerky futile attempts to reach the floor.

Cousin June floated by with Manuel and, judging by the teeth and ho, ho, ha, ha, ha's, Greek had met Greek. All around the edge of the floor children danced and rassled and fought and banged into the dancers. Women danced together and sometimes three people would dance together screaming with laughter at their originality. Everyone was having a wonderful time.

Suddenly the music stopped, and the entire crowd and the orchestra surged to one end of the hall where the bad Indian girl and some farmer's wife were pulling each other's hair and smearing each other with richly appropriate epithets such as "Cross-eyed daughter of an egg-raising whore" and "Dirty clam-eating black bitch." The Indian girl was the drunkest so she was put out, the music began again and we danced until supper. Supper, served at midnight in the basement on long tables, consisted of hamburgers, potato salad, cake, ice cream and big mugs of coffee.

After supper I danced twice with Bob and once with a very drunk sailor who called me "a very poor shport" because I refused to let him lower me by the ankles from a window where a pal of his waited with a drink some twenty feet below. "I'm dishappointed in you," he said, blinking and swaying. "I'm sorry," I said, "but my husband is very strict with me." Then I saw Birdie's mother across the room, coiled and waiting for her next partner. I grabbed the sailor and gave him a little shove in her direction. "That little lady over there in the pink dress is just the person you're looking for," I told him. "I know that she'd love to get a drink from your pal."

Whether Mother was lowered out the window to Pal I never knew, for we left while they were still dancing, but I can picture her flipping out the window and up and down between Bluejacket and Pal like a yo-yo on a string. So much pep!

As we started the long drive home Mrs. Hicks began tossing us back bits of gossip picked up at the dance. The last I remember was "and Mrs. Cartwright's son's wife, Helen, lost her mother and then a tube and both ovaries. . . ." I wondered drowsily if she had lost them at the dance and then fell asleep with my nose buried in Bob's shoulder which smelled of cheap Indian perfume.

The most outstanding of all (five) of the social events to which we were invited during our stay in the mountains, was Mrs. Kettle's birthday dinner. We were the only outsiders asked, which was a singular honor; it was a Kettle gathering on Kettle soil and our first introduction to them en masse.

One very hot July morning Mr. Kettle and Elwin drove up to the house, to borrow some egg mash. I invited them in for a cup of coffee but Paw refused. He said, "Thure would like to but you thee itth Mawth birthday tomorrow and Elwin and I are trying to get a little help with the haying tho we will be free to thelebrate. Do you thuppose that Bob could thpare uth a little time?" I said, "Bob is working on the water system and I know he won't be able to help. What about the other boys? Can't they help?" Elwin said, "Haw, haw, haw! Help with the haying? They gotta work in the woods." Paw said, "That's the WAY IT GOETH. THE BOYTH WON'T help and the old lady can't do it all alone." I said, "I thought you and Elwin were going to do the haying." Elwin said defensively, "I gotta get my car fixed before Saturday." Paw said, "Mr. Olson cut our hay last year and Maw raked and stacked it—I wath havin' turrible trouble with my back—but theein ath how tomorra is Mawth birthday we thought we would give a little thupper party for her and we want you folkth to come but she'll be tho buthy she won't have a chance to help with the hayin'."

I repeated again firmly, "I know Bob won't have time to help with the haying."

"Well, thatth too bad," said Paw. "But don't forgit to come to the thupper."

"I won't," I said. "What time?"

"Oh, 'bout four-thirty," said Paw. "And by the way, everybody ith bringin' a little somthin' to the party. You know thort of picnic thtyle."

"I'll bring the birthday cake," I said, which evidently satisfied Paw for he and Elwin took their egg mash and left. (They came up again in September to solicit help with that same hay.)

I was not an expert cake maker but decided to make up in size what it would undoubtedly lack in quality. None of my recipe books contained any directions for anything over two layers so I doubled and tripled and before evening I had the great heavy square foundation layer done.

The next day dawned hazy but warm—a perfect day for a party and a perfect morning to bake the rest of the cake, which by noon was finished and frosted pink, white and blue. It was poisonous but festive looking, and as a final corny touch I lettered HAPPY BIRTHDAY on one side in red cinnamon drops and stuck an American flag on the top. Bob drove to the Crossroads for a pair of silk stockings and a birthday card and on the way back picked up Birdie to stay with the baby. At four o'clock on the dot we loaded the cake and set off. The farther down the mountain we drove the hotter it grew until, by the time we turned into Kettles' yard, it was stifling. The yard was seething with cars, as all of the fifteen children, their husbands, wives, children and in-laws were there. Mrs. Kettle was bright-eyed in a clean pink housedress which was missing the top two buttons in the front so that as she greeted us I noticed that her large white breasts bobbed to the surface like dumplings in a stew. I handed her our present and she kissed me, unwrapped the package and said, "Well thank God, I don't have to wear them damn lizzle stockins no more."

She directed me to put the cake on the front porch where we were to eat. I carried it through the dark front hall and out the never used front door and set it with a thump on the long table of boards on saw horses which had been set up and covered with a pink crepe paper tablecloth anchored with rocks. Almost all of the chicken droppings had been scraped from the floor and railing and the lilac bushes cast cool shadows across the table. Two of the flypaper curls had been suspended from the ceiling and already they were dotted with flies and swayed and quivered as more hit and stuck. The little flag on my cake drooped as the icing grew warm and soft and the HAPPY BIRTHDAY seemed about to slide off at one end, so I moved it farther into the shade before returning to the hot kitchen and the mobs of guests. As I left the porch I noted that, whether we had counted on it or not, the outhouse was going to be noticeably present at supper; so I suggested when I went to the kitchen that someone put up some sort of temporary door across its front. Mrs. Kettle said, "It's a good idea but I don't think you'll get any of the men to do it for you. Go shout at 'em, they're on the back porch."

I shouted, but the men were busy drinking blackberry wine, moonshine, dandelion wine and beer and comparing notes on guns and hunting, and the only answer I got to my request was a look of annoyance at my interruption from Bob. We women drank coffee while we made potato salad, cut the meatloaf, opened pickles and sliced bread. It was stifling in the crowded kitchen but we had to build up the fire to bake the rolls and boil the fresh coffee. I opened the door to the parlor, hoping to release a cross current of air, but the children immediately crowded into this new area, so Mrs. Kettle got up, firmly shooed them out and locked the door. The sanctity of the parlor was not threatened even on her birthday. The children raced from the kitchen through the hall out the front door, off the porch, around the house, up on the back porch and through the kitchen again. They dodged under hot platters, stuck their fingers in the jam and got whacked, rolled on the floor with the dogs, came to get their panties unbuttoned or buttoned, and tattled. The big rocker by the stove held a succession of mothers suckling babies, while the other chairs supported the ample behinds of helpers and cooks. The talk was about sex, logging, sex, cooking, sex, sewing, sex, salaries, sex, fits, and just plain sex. Every four letter word I had ever heard and many I had never heard were employed singly and in unusual combinations. In anger or only slight irritation the children were called such unbelievably filthy names, that in comparison "dirty little bugger," "Christly sonofabitch" and "stinking bastard" seemed terms of affection and endearment. Never have I encountered a group that needed "heart medicine" more, unless perhaps it was the men who, Bob told me, when they weren't slipping from the porch to siphon gas or lift any loose tools, radiator caps, or gearshift knobs from each other's cars, were loudly relating incidents to exhibit their sex prowess either before or after marriage. Mrs. Kettle laughed occasionally but she was plainly embarrassed, as was I. We cut up the potatoes and hard-boiled eggs and talked about canning and were both very relieved when the enormous granite coffee pot finally began to boil and it was time to set the table and carry out the food.

Everyone on the back porch was by that time quite drunk and their voices had become loud and a little quarrelsome. A small boy, a little bully who had been causing a great deal of unhappiness among the younger children, came screaming up from the barnyard where he had been chased but unfortunately not eaten by an old sow. Maw sent Elwin to fetch Mr. Kettle from the barn and we all filed out to the porch for supper.

When everyone was seated, I suggested that we sing *Happy Birthday*. This was a mistake, I realized too late, for they followed *Happy Birthday* with *Show Me The Way to Go Home* and the dirty version of *Little Red Wing*. Maw interrupted this last by banging on the table with her fork and demanding quiet. Everyone cheered and said, "At's tellin' 'em Maw!" and began passing food. Elwin had said that Paw would be right up, but we were well into the ice cream before he finally appeared. To protect him from the flies he had put on and pulled far down over his eyes, a black straw hat of Mrs. Kettle's. From its crown bobbed a large pink rose. To shield him from draughts he had thrust his arms into a jade green knitted coat dress with a full pleated skirt. The skirt must have interfered with the milking for he had taken a deep hem in it with horse blanket pins, giving the effect of a ballet skirt. He also wore a black work shirt, dungarees, and hip boots and was smoking a cigar. Choking in an effort to control my laughter until the others had seen the joke, I glanced at Bob, but he shook his head at me warningly. Amazed, I looked around the table—not one soul seemed to think there was anything unusual in Paw's dress. He clumped around the table, smiling happily at everyone, climbed into the chair beside Mrs. Kettle by swinging one manurey boot carelessly over that end of the table, and she said only, "You've took so long with the milking your coffee's cold, Paw. Helen, this here's the knitted dress Myrt sent me from New York. Ugly color ain't it?"

After supper the men returned to the back porch, we

women put the food away and washed the dishes, and Bob and I left just as a sizable fight started over who had stolen what from whose car. The cake was a great success.

We seldom went to the movies for two very sound reasons: 1. It meant driving from seventy to one hundred miles. 2. We had to get up at four o'clock the next morning. In addition to these, Bob complained that his legs were too long for comfort in any theatre seat; he always slept through everything but the news; and I became so biased that no matter how melodramatic the plot, I watched only to see if the heroine did any work or if she seemed to have all of the conveniences of modern life. If she didn't work and seemed to have plenty of opportunity to take big hot steamy fragrant baths, I lost all interest in the plot. Under such circumstances who gave a damn who got the man.

I will go to a fair any time. A fair to me is the ultimate in entertainment. Something of interest for everyone, all the time in the world to enjoy it, and so many delightful smells—popcorn, coffee, molasses candy and wet sawdust. For years our family had made an all-day visit to the Western Washington Fair at Puyallup but we treated the trip purely as entertainment. We had no active part in it and we paid no attention to prize ribbons other than in the flower department where Mother forced us to wait while she bought bulbs and seeds.

I entered our County Fair with such an entirely different viewpoint that it was as if I had never been to a fair before. The County Fair was held near Town in late September. Town was in an odd little dry belt which meandered through that soaking wet country so there was never any danger of the fair being spoiled by rain, as it so often was at Puyallup, and by the opening day the dust was ankle deep and had to be hosed down every morning giving a pleasant spring-rain smell to pathways. We attended the fair in true farmer style. I packed dozens of diapers, several bottles, blankets and pillows so the baby could take a nap in the truck, mosquito netting to keep off the flies, extra old shoes for the trip through the animal pens, and we were on the highway by eight in the morning.

It was chilly and foggy in the mountains, but Town was flooded with bright sunshine and whipped by salty breezes. The water was deep blue and lacy with whitecaps, the mountains were pale blue and white with sharp outlines, the hills were soft golden yellow and the old red courthouse gazed down on the town with the dignity and benignancy of a cathedral. Despite the early hour there were many other cars and trucks on the road and when we turned into the fairgrounds we found as much activity as though it were late afternoon. We parked the truck under a large Balm of Gilead tree near the restaurant, I divested Anne of her knitted leggings, put on my old shoes and we were ready. The first thing, of course, was to have a cup of coffee and some fresh homemade doughnuts. At stools along the counter on either side of us were other farmers and their wives and children. The other babies were being given sips of coffee and bites of the hot greasy doughnuts and I felt hard and niggardly as I ate and drank with Anne's round blue eyes following each sip and bite like a little puppy. After the coffee we went to the poultry show—of course, we thought that some of the White Leghorns that had won blue ribbons weren't a bit better than our own hens. Bob was much interested in the Australorps—black chickens which were supposed to lay as well as the White Leghorn and yet be as fine a table bird as the Rhode Island Red or Barred Rock. It sounded too good to be true but the farmer who exhibited swore that they were the most remarkable discovery since electricity. From the poultry houses we went to the pig pens which were more pleasant early in the morning; then to the sheep, goats and rabbits. By this time Anne was getting sleepy so I took her back to the car and left Bob with instructions to meet me at the restaurant at noon. I fastened Anne to the seat of the car with a blanket under and over her and pinned to the upholstery, stretched the mosquito netting from the back of the seat to the steering wheel and she immediately and obligingly went to sleep. I spent the rest of the morning in the canning and fancywork exhibits—I felt a surge of pride when I saw that Birdie Hicks had won blue ribbons on all of her canning and that Mrs. Kettle's quilts were prominently displayed. There were some really beautiful hooked and braided rugs and some perfectly hideous ones made of gunny sacks fringed and coated with yarn forced through the mesh and tied in knots. I remember seeing instructions for these in one of the farm magazines. There was a pen and ink drawing of the rug, showing it with a nap of about eight inches. Underneath it said "There is beauty in even the common grain sack." There is, too, but not when it has been disfigured with yarny knots. There were other ingenious uses for everyday objects. There was a Sears, Roebuck catalogue painfully twisted and shellacked and tied with a red cord. The white card beneath it said, "An inexpensive doorstop"—Ethalynne Weatherby. I made a mental note to ask someone if that was "our Mrs. Weatherby." There were catsup bottles made into bud vases, clothespins decorated with crepe paper butterflies for use as curtain holdbackers, crocheted bags for silverware, bouquets of crepe paper and velvet flowers, an enormous funeral set piece of white organdy gardenias and dark green oilcloth leaves with REST IN PEACE spelled out in white pipe cleaners, embroidered pictures, burned wood match boxes, and fancy pillows by the hundreds. The pillows embraced every sentiment from

FRANKY AND JOHNNY WERE LOVERS in black beads on a cerise satin background to the Twenty-Third Psalm in white on black velvet. It was an impressive exhibit of what loneliness can do to people.

23

Put Out That Match!

EVEN with the continual rain, July, August, September and even October were bad fire months in the mountains. If you were unfortunate enough to live on a ranch near the Kettles, any month was dangerous. It was said that the Kettles set the original peat fires in the valleys and that one summer, Paw, to save himself the effort of mowing the lawn, set fire to the grass and burned off the front porch. The Kettles burned brush any old time of year and if the brush fire got away from them and burned five or ten acres of someone's timber, that was too bad.

In the Northwest, particularly in regions near the salt water, the underbrush is rank and grows with tropical rapidity. Salmon berry bushes, wild syringa, thimble berry and alder grow six to ten feet in a season and make land clearing a yearly task. When nothing is done to keep back this underbrush, the old canes remain tinder dry underneath and in between the new growth. This makes danger from fire a perennial thing.

It was late October of the second year when Elwin Kettle drove excitedly into our yard one morning to tell us that their barn was on fire. He said, "Paw filled the hay mow with wet hay and the darned stuff's combusted spontaneous. The barn's burnin' and it's set the brush in the ravine on fire, too. Maw said to warn you that it'll probably come up this way because the wind's from the south." I was frightened and ran out to the chicken house to fetch Bob. Bob said, "Oh, I don't think it will amount to anything," but he took the truck and followed Elwin back to the ranch. Bob did not return until noon. When he did come home he was black from head to foot and very angry. He said that he was quite sure Paw had set the barn on fire to solve the manure situation; that the barn had burned, but the fire had left the Kettle ranch and was sweeping up the draw with terrible velocity.

He said that if the fire crossed the road, our ranch would go. I asked if the Kettles were helping fight the fire and Bob said, "They helped only as long as the fire was on their land. The minute it crossed their boundary fence they went back to the house on the pretext of protecting the house and outbuildings. When I went to get the truck I saw them all on the back porch, drinking home brew and laughing and talking."

Bob told me to spend the afternoon hosing off our roof and soaking the dry orchard grass and preparing to feed the fire fighters. He left to get help.

All afternoon cars chugged up our road and then scooted down toward the Kettles'. Bob had spread the alarm and the farmers were answering the call for help. I learned later that Bob had gone even farther—he had asked Maxwell Ford Jefferson, the moonshiner, to help and Jeff passed the word on to all his good customers. I made gallons of coffee and hundreds of sandwiches and at about four the fighters began filing wearily in, black faced and dirty. They ate sandwiches, drank coffee, and cursed the Kettles. The smoke rolled up the draw and obscured the sun and stung my eyes while I put out the lanterns and fed the chickens. Each batch of fighters brought more frightening news. "The fire was almost to the road in several places." "The wind was getting stronger."

"Once the fire hit the timber on our place we would have to run for our lives."

About seven o'clock Max Jefferson knocked at the back door. He was all alone. He said in his soft liquidy voice, "Just thought I'd drop up and tell you not to worry, honey. I've got the biggest bunch of drunkards in the United States out fightin' that fire 'cause they think my still's up this away." Jeff was a tall, tobacco-colored, lithe man with old world courtesy of manner, a southern drawl and light yellow eyes which saw in the dark and spotted every exit in a room before he crossed the threshold. He drank his coffee and ate his sandwiches, tilted back in a kitchen chair, occasionally fingering the gun in his coat pocket and flashing his big white teeth. After he had finished his coffee he said he had something for me in his car and went out and brought in a ten gallon keg of whiskey. He set it down carefully in the middle of the kitchen. "That theah is fine whiskey," he said. "It's been hangin' in a tree and should be smooth as oil now." Before I could thank him he had slipped out the door and was gone.

I put the baby to bed, then went out to the front porch to watch the fire. It was only about two city blocks from the house and the snapping and crackling were fearful sounds. The fire fighters reported that they had kept it out of the heavy timber down by the water pump and that the wind was dying down. I wandered aimlessly around after each batch of fighters had eaten and left, trying to choose what I would take in case we had to run for it. It narrowed down finally to the baby, the keg of whiskey and the animals. I put my sorority pin and my high school graduation ring in my purse, my silver in the didy bag and we were ready.

About eight o'clock Mrs. Hicks arrived with two chocolate cakes, two pounds of butter and two quarts of cream. The smoke had grown heavier with the lessening of the wind, but by closing all of the doors and windows I was able to keep most of it out of the house. A group of fighters arrived simultaneously with Mrs. Hicks and we had first to feed them before we could talk. Then we poured out cups of coffee and sank down to discuss the fire. I noticed with pleasure that in the excitement Mrs. Hicks had pulled the keg of whiskey over by the stove, had placed a pillow on it and was sitting there primly drinking her coffee.

Bob came excitedly in, grimy beyond recognition, and demanded help in soaking and loading in the truck the hundreds of feed sacks we had stored in the loft of the tool house. As we worked he told me that Jeff had the Kettles working at the point of his gun. He said that poor Maw Kettle was so humiliated about the whole thing that she had come to him in tears to ask if there wasn't something she could do. She had made coffee and fried bread and had fed many of the fire fighters.

After Bob had driven away with the wet sacks and ten milk cans of water, I went out to see how the heifer was reacting. She was in her stall, standing quietly but not eating. She had a nice new little barn beyond the pig houses, the farthest from the house of any of the buildings, so I led her back to the house and tied her to the cherry tree. Sport and the puppy trailed me everywhere, whining and begging me to explain the smoke and excitement. Sport's method of eliciting understanding and comfort from me was to shake hands. Every time I hesitated during the whole of that long, dreadful day, I would look down and there would be Sport, pale green eyes spilling over with love, offering me his large soft paw. By ten o'clock his paw and my hand were calloused and I was tired and scared so I yelled at him, "Oh, Sport, don't be such a bore!" Whereupon he gave me a look which said, "Here I am willing to die for you, trying to comfort you in your hour of need, and you speak to me like that." I gave him six chocolate creams and put him behind the stove where he remained for the rest of the night, shivering occasionally and whining steadily. Birdie Hicks left about ten-thirty with my promise that if anything happened I would stay with her indefinitely.

After she had gone I sat down by the stove with a magazine and I must have dozed for I was awakened by loud shouting down by the road. "This is it," I thought. "Let me see, should I wake the baby now, or wait until Bob comes." I was shivering and so I put some more wood on Stove thinking what a useless gesture that was when the whole house would soon be in flames. I put the diaper bag on the whiskey keg, laid my purse on the diaper bag, glanced out the window to see if the heifer was still tied to the cherry tree, and we were ready. I could hear voices coming up the lane and my heart beat wildly. "Where was Bob? Had the truck been burned?" The back door opened and Jeff came in looking like something from a minstrel show. He said, "It's all O.K. honey, the fire's in the burn and it's fixin' to rain in about ten minutes. Better start slicin' ham and fryin' eggs and boilin' coffee—there's an awful hungry mob on the way up."

I rushed and untied the heifer and led her back to her stall. The smoke was so thick it was like a heavy fog but there was a cool wet feeling in the air.

I fried eggs and ham and bacon and made toast and coffee until five the next morning. I also listened to the details of that and every other fire on the coast, since the days of the first settlers. Paw Kettle and the boys were very much in evidence, behaving like heroes instead of the guilty perpetrators of the disaster, and with the camaraderie produced by success and my keg of whiskey every one else adopted the same attitude, getting Paw to tell over and over when he first smelled smoke, what he did, if he lost any livestock, and so forth. I was very grateful when Paw at last brought the gathering to an abrupt close by pointing out that in the early days when a farmer needed a new barn all of his friends and neighbors got together and helped him build it. Being Paw, he couldn't let it go at that—he had to add, "And I heard tell that eath NEIGHBOR WOULD BRING thomething—thay, one would bring nailth—another the two by fourth, another THINGLE BOLTH . . ." Before he could finish this big lie, everyone but Bob and Jeff had eased out the door.

24

You Win

THE NEXT DAY it rained hard and all that remained of the fire were a hillside of glistening black stumps, occasional sharp cracks, wispy smoke and an acrid smell. Bob slept until noon, then arose groggily, gulped a cup of coffee and drove to Town to see about the lumber and millwork for an addition to the chicken house. He planned to brood twenty-five hundred baby chicks in the spring, which would give us, even with heavy culling of the old flock, two thousand laying hens the following fall. He also planned to buy a cow, scheduled to come fresh in March, one hundred baby turkeys and five young pigs. He was also going to see a farmer in the West Valley who had a light plant for sale.

Our future prospects were very good but my enthusiasm was at a low ebb. I was overtired by the fire and insufficient sleep and even the magic words "electric lights" couldn't dispel the gloom in my outlook. My life on the ranch had reached some sort of climax and it was the aftermath which worried me. We were just about to go into another long, dreary winter and I felt harried and uncertain as though I were boarding a steamer with no passport and no luggage. I was leaning on the drainboard of the sink, staring moodily out of the window at the driving rain and drooling eave troughs when a figure in a long billowing white dress and without a coat or hat came loping into the yard. My reactions were delayed that day, for I pensively watched while this odd person galloped through the rustic gate, her long dress flapping wetly at her legs, dashed over to the cherry tree, picked up a wooden duck of Anne's, cuddled it to her bosom and began dancing around and around the cherry tree. Suddenly I came to. Something was not right. The woman had a shorn head and was obviously crazy. "This is all I need," I remember thinking frantically. "This finishes it." I knew that there was not a lock on the house which would keep this woman out if she made up her mind she wanted to come in. I was so frightened that I was probably right on a mental plane with my visitor. All the blood had drained from my body with a rush, leaving me perfectly flaccid. I thought, "I must lock the doors. That will give me time." I got to the kitchen door somehow and locked it. The kitchen door had one of those ridiculous locks which is a small black box with a tiny lever on the top; an angleworm could have forced it. My fumblings at the back door attracted the woman's attention and she came running across the yard crab fashion and peered in the kitchen window. Her eyes rolled in her head like marbles and she laughed wildly as she hurried from window to window playing peek-a-boo with me. I was leaning on the stove whimpering, "What *will* I do? What will I do?" when I remembered the open screenless window in the baby's room where she was taking a nap. I grabbed the stove lifter and started for the baby's room which was across the living room on the other side of the house. I couldn't hurry. I felt as if I were wading in water up to my waist. It took every ounce of strength I could muster to push one foot ahead of the other. When I finally reached the baby's room the woman was there already, her head and shoulders in the open window. I brandished the lifter, feebly croaking, "Get out of here. Go away!" She stopped rolling her eyes and laughing and looked at me. Her face crumpled; she looked as if she were going to cry. Then, slowly she retreated from the window, turned and went loping off down through the orchard toward the Kettles.

I watched until she had rounded the first bend in the road, then I sank weakly down on the bed and thought, "Now I'm going to have hysterics. I'm going to explode just like a sky rocket." But the baby awoke then and, in the adorable way of babies, was overjoyed at seeing me and jabbered and held out her arms and that was that.

Bob came home an hour or so later and he had new magazines and cigarettes and candy and a truckload of lumber. He was bubbling over with ideas and plans for the ranch and so I waited until after supper to tell him about the crazy woman. I didn't even try to convey my terror because I knew by then that Bob and I were poles apart as far as emotions were concerned. I knew before I had finished the story what he would say and he said it. "Why didn't you get out a gun? Always remember that with a loaded gun in your hands you have the upper hand of anything." That was true for Bob but not for me. With a loaded gun in my hands the gun had the upper hand and besides, if you are the kind of person who grabs a stove lifter instead of a gun when danger is at hand, you are that kind of person and you have to face it.

After supper Bob and Anne and I drove to the Kettles' to find out if they had seen the crazy woman. Mrs. Kettle was not at all perturbed. She said, "Oh, she was in here this afternoon. She's the loony sister of a woman down in the West Valley. Mostly she is shut up in an institute but whenever they can save the money, they have her over for a visit. She won't hurt nobody—she's just loony."

"Where is she now?" Bob asked.

"Why, she's down at the Larsens' farm," Mrs. Kettle said.

"They phoned for the sheriff and he'll probably send her back to the institute."

"Oh, root-ta-ta-toot, a root-ta-toot—we are the girls of the institute!" Bob sang as we drove home. I thought, "That valley woman must be even more isolated than I, if she's that desperate for companionship."

The next morning the storm had increased in tempo and there were pools on the floor beneath the open windows and the rain had oozed under the doors during the night. The baby's washing festooned the kitchen once more and Stove had his back up and refused to digest the fuel and became constipated with ashes. Swathed in oilskins Bob began work on the new chicken house. After I had finished the lunch dishes and put the baby down for her nap, I donned my rain clothes and went out to help him. I peeled stringers and measured two by fours. I ran and got the hammer, the saw, the level, the rule. I was more skilled than last year but I had lost all of my drive. Bob said, "Would you like to start splitting the shakes for the roof?" I loved to split shakes and ordinarily I would have been enthusiastic but all I could manage then, was a weary, "All right."

At dinner Bob shattered all precedent by suggesting that we drive to Town to a movie. He said he had asked Max Jefferson to stay with the baby, so we hurried with the dinner dishes and the chores, dressed, stoked the fire and then sat down to wait. When fifteen minutes had elapsed, we realized that we would have to see the second feature anyway, so took off our coats and Bob lit us each a cigarette. We sat and smoked and were self-conscious with each other. Bob studied the burnt end of the match he had used for our cigarettes, turning it around slowly in his lean brown fingers. I watched him and the clock ticked off the minutes. "Do you think he's coming?" I asked at last. "Oh, he'll be along," Bob said, not looking at me. I thought, "Heavens, we act like neighbors who suddenly find themselves in a hotel bedroom together." The clock ticked on. Bob said, "Did you order those hinges?" I said, "Oh, I forgot," and jumped up guiltily and started for the cupboard where we kept the catalogues. Bob said, "Don't bother now," and got up and put his cigarette in the stove. I came back and sat down. Bob was poking aimlessly at the fire, his back still toward me. I lit another cigarette and reflected, "Husband and wife teamwork is just fine except when it reaches a point where the husband is more conscious of the weight his wife's shoulder carries than of the shoulder itself." I said, "I don't think that Jeff is coming."

Bob said, "I guess not," and sat down and lit another cigarette.

The clock ticked on and after a while we went to bed.

October eased out and November slid into its place. We got up in the morning to the persistent steady thumping of rain on the roof and we lay in bed at night and listened to the persistent steady thumping of rain on the roof.

One Saturday I went to town but when I came home I knew it would be the last time that winter—driving so many many miles in the cold and rain in a truck was too hard on the baby.

Winter had come again but there was no excitement in the knowledge. No visions of exhilarating battles with piling snows; no shutting out of roaring blizzard; no sudden dramatic changes in the weather. It was the regular, gradual closing in of the mountains with rain, rain, rain, moaning winds and loneliness.

Just before Christmas Bob made an unexplained trip to Seattle. He was gone three days. The second day I ran out of kerosene and we had the worst storm the mountains could manage on such short notice. The wind screamed around the house like a banshee, the trees beat their breasts and moaned, and the night threw her musty cape over us at a little after four in the afternoon. I lit candles which sputtered feebly, poked up the stove, put the baby on the floor beside Sport and went out to do the chores. The animals all seemed apprehensive and hinted broadly that they would prefer to spend the night in the house with me, but I was firm and fed and watered them, then bolted their various houses and barns against the wicked wind. When I returned to the house, bedraggled and dripping, I found that the back door had blown open, the candles had blown out and the baby was howling. With trembling hands I relit the candles, poked up Stove and fixed the dinner. At five-thirty sharp we were all in bed. In the same bed. Anne and I under the covers at the top. Sport the puppy, and the kittens at the bottom.

I left the candles lit although it was hard to choose between being burned alive and being scared to death.

The next morning dawned at last gray and wild. Bob came home at three, looking and smelling deliciously urban. He was so cheery that I hesitated about telling him that I had again forgotten to order kerosene and in the ensuing excitement I forgot all about it, and we spent another evening by meager candlelight. Bob had news. Wonderful news which he told me as he changed his clothes. He had gone to Seattle on the advice of the Town banker, to look at a chicken ranch which was for sale. He hadn't said anything to me because he didn't want to disappoint me if the ranch had proved a poor buy. But it hadn't. The buildings were sound and well built although in need of repairs and paint, the price was stiff but not unreasonable and the house was modern. In addition to all that, the Town banker had a buyer for our place. One with cash.

The storm continued, the candles blew out, the stove remained lukewarm but I didn't even notice. I coasted around the house propelled by visions of linoleum floors, bathtubs, electric stoves and flushing toilets. It seemed to me that from now on life was going to be pure joy. After dinner we sat at the kitchen table and by the light of the sputtering candles figured assets and liabilities. At least Bob did. I was busy figuring how many hours a day I would save by having modern conveniences. I said to Bob, "I suppose that with lights in the chicken houses and running water and things we wouldn't have to get up until about seven or half past." Bob was busy figuring. He said, "Huh?" I repeated, "I imagine that with lights and running water in the chicken houses we wouldn't have to get up until about seven or half past."

"Oh, that won't make any difference," Bob said. "Chickens have to be fed anyway and the earlier you feed 'em the sooner they start to lay." Which just goes to show that a man in the chicken business is not his own boss at all. The hen is the boss.

THE END